TIMBER FRAMES
DESIGNING YOUR CUSTOM HOME

Rock Island Public Library
401 - 19th Street
Rock Island, IL 61201-8143

FEB 2009

DATE DUE

GAYLORD #3522PI Printed in USA

TIMBER FRAMES
DESIGNING YOUR CUSTOM HOME

BY JEREMY BONIN

with Rebecca Sandiford

THE HELICONIA PRESS

THE HELICONIA PRESS

The Heliconia Press, Inc. 1576 Beachburg Road, Beachburg, Ontario K0J 1C0 Canada
www.helipress.com

Text and Architectural Design © Davis Frame Company
Photographs © Davis Frame Company
Illustrations by: Jeremy Bonin
Computer Renderings by: Christopher Timberlake

All rights reserved. No part of this book, including photos, architectural drawings and illustrations, may be reproduced in any form, or by any electronic, mechanical, digital, or other means, without permission in writing from the publisher.

Written by: Jeremy Bonin with Rebecca Sandiford
Edited by: Rebecca Sandiford, Lise Wilson
Photographs by Rich Frutchey, Davis Frame Company and Marcy Bahara.
Design and Layout by: Vicki Veenstra

Library and Archives Canada Cataloguing in Publication

Bonin, Jeremy, 1972-

    Timber frames : designing your custom house / by Jeremy Bonin ; with Rebecca Sandiford.

ISBN 978-1-896980-34-8

    I. Wooden-frame houses--Design and construction. 2. Framing (Building). I. Sandiford, Rebecca, 1973- II. Title.

TH4818.W6B65 2007   694'.2   C2007-903149-8

ACKNOWLEDGMENTS

It seems fitting that I am penning my words of gratitude in a nearly finished timber frame room. At this moment, the book is much like the room; both are just waiting for those finishing touches that will make them really come to life. Writing this book, which has become so much more than I initially imagined, has mirrored a part of what I find so appealing about architecture. A design always goes on to become something more than what was first envisioned, simply by virtue of the people involved.

    I would like to give very special thanks to my father and John Deming, the General Manager at the Davis Frame Company. My father, in the span of a two-hour car ride, convinced me that moving my family, leaving a good position and making a leap was worth it—if it was for something I really wanted. As usual, Pops was right. John Deming urged me to consider the role of design team lead at the terrific company he worked for, and he also turned out to be right. Without these two people, this journey and this book would never have been realized.

    Many thanks to Jeff Davis and Rick Bascom of the Davis Frame Company for always answering my many questions and their support for this book. Their enthusiasm for the work they do is truly infectious, and the company they have built together is a huge part of what makes my job a pleasure.

    There were several people who were forced to put up with me while this book was being written. To my wife and daughters, thank you for tolerating my agitation as I discovered exactly how hard it is to write a book and the resulting general craziness in our lives. To my staff and coworkers, thank you for generously accommodating my moments of distraction and my erratic schedule; your support means a lot to me.

    I drew inspiration for parts of this book from architect and author Sarah Susanka whose values and philosophy of quality over quantity apply so beautifully to the art and craft of timber framing. I was also inspired by an architect, author and professor whom I had the privilege of studying under in college, Professor Glen Wiggins. His sincerity, enthusiasm and willingness to speak his mind impressed me then and stay with me now. Thank you.

    I am grateful to Rebecca Sandiford, who edited this book and helped to organize my writing and clarify my thoughts; and to Ken Whiting and the rest of The Heliconia Press gang for their great work and being such a pleasure to work with. Many thanks also to Jackie Lampiasi of Davis Frame Co. for her assistance with the graphical content.

    Finally, I'd like to thank you, the reader, because by picking up this book and being curious about its contents, you make it all worthwhile. For what good is any work without those who will engage with it?

TABLE OF CONTENTS

ACKNOWLEDGMENTS	7
INTRODUCTION	11
FOREWORD	13

CHAPTER 1: INTRODUCTION TO TIMBER FRAMING — 15

- WHY A TIMBER FRAME HOME? — 15
- A BRIEF HISTORY — 17
- A PRIMER — 21
  - Frame Terminology — 21
  - Types of Frames — 22
  - Types of Joinery — 28
  - Types of Wood — 31
  - Timber Frames and Insulated Panels — 31
  - Types of Construction — 32
- CHOOSING TIMBER FRAME AND JOINERY TYPES — 37

CHAPTER 2: GETTING STARTED — 39

- DOING THE GROUNDWORK: EXPLORATION — 39
- DEFINING YOUR VISION — 45
- IMPORTANT CONSIDERATIONS — 46
  - Fundamental Considerations — 46
  - Elemental Considerations — 47
- CREATING AN OUTLINE — 51
  - Guiding Questions — 52

CHAPTER 3: PRE-DESIGN — 73

- KEEPING PERSPECTIVE ON THE DESIGN PROCESS — 73
- PHASES OF DESIGN DEVELOPMENT — 74
- ABOUT PRE-DESIGN — 74
- UNDERSTANDING YOUR SITE — 74
  - Site Access — 77
  - Utilities — 77
- REGULATORY GUIDELINES — 80
  - Zoning Regulations — 80
  - Building Codes — 83
  - Specialty Regulations — 83
- BUILDING SYSTEMS — 84
  - Fuel or Energy Source — 84
  - Heating and Cooling Systems — 87
  - Building Shells — 90
  - Lighting and Power — 96
- PROJECT SCOPE AND RESPONSIBILITIES — 99
- PROJECT SCHEDULES — 99
- PROJECT BUDGET AND CONTINGENCIES — 99

CHAPTER 4: REFINING THE DESIGN — 101

- SCHEMATIC DESIGNS — 101
  - Nurturing Ideas — 102
  - First Steps — 102
  - Intermediate Steps — 105
  - Final Steps — 106
  - Timber Frame Planning — 106
- DESIGN DEVELOPMENT — 107
  - First Steps — 107
  - Intermediate Steps — 109
  - Final Steps — 110

CHAPTER 5: CONSTRUCTION DOCUMENTS — 113

- ARCHITECTURAL DRAWINGS — 114
  - Cover Sheet — 114
  - Floor Plans — 114
  - Elevations — 115
  - Building Sections — 115
  - Foundation Plans — 116
  - Sections and Details — 116
  - Schedules — 116
  - Frame Drawings — 117
  - Panel Drawings — 117

AFTERWORD — 119

GLOSSARY — 120

INTRODUCTION

Designing a home is an evolutionary—at times revolutionary—process. A custom-designed home is as unique as the people who will inhabit it, whether the home is built as a primary home, vacation home or a home to retire to; using modern conventional methods, timber-framing or a bit of both. The design process is what enables that singular home to emerge.

The process of designing a house occurs in stages, with each new stage building on the actions taken and the decisions made in the preceding ones. While there are benchmarks that can be used to define the end of one stage and the beginning of the next, it's important to remember that the most successful design processes are iterative or incremental in nature, often forcing the re-evaluation of choices and preferences identified in earlier stages. To wax poetic, you can think of these iterations or stages as being a little like life itself; if we allow the challenges that we encounter in the design process to be tools of refinement, we make room for maturation in the design that will be evident in the final product, a home for people to inhabit and enjoy.

This book was written to be the definitive primer for anyone interested in the adventure of designing a custom timber frame home. It will guide you through the process of designing a timber frame home, including a few exercises that you should complete before you start, and then through the different stages of design refinement. You'll learn the terminology for different types of frames, and you'll find valuable information about the latest innovations that are important to consider. Along the way, you'll find helpful tips, suggestions and tricks that will help you make decisions and prevent unnecessary headaches down the road. Let's get started!

FOREWORD

In our many decades in the construction industry—almost a century between the two of us—we have discovered a simple truth: that any building that is thoughtfully designed and well-built is a gift to future generations.

Beautiful, comfortable homes that transcend the passing trends never go out of style; and furthermore, they can be easily adapted for new uses over time. Energy-efficient, durably built and structurally robust homes, made with renewable and recyclable materials, are a respectful use of the planet's precious resources and considerate of those who will come after us. Timber frame homes offer all of these advantages. For these and many other reasons, we love building them.

This book has been written to support you through the process of imagining, planning and designing your timber frame dream home. Sometimes custom designs come together almost of their own accord; but in most cases the design process will present many interesting challenges. With so many factors to balance and consider—among them your needs, wants, aesthetical preferences, budget, site restrictions and the laws of physics—it will demand patience, creativity and a bit of hard work. We can assure you, however, that it will be worth it.

Timber frames allow you to create every type of space your home might need: from the grand and expansive to the small and cozy; from public gathering areas to private, intimate spaces; from recreational to functional. For every type of space, timber frames lend a sense of warmth, natural beauty, symmetry and simplicity. Timber frames provide a sense of security too. We have seen firsthand how well timber frames stand against hurricane force winds, unusually heavy snow accumulations, and the rock and roll of earthquakes.

We feel that by designing and building energy-efficient timber frame homes, we are honoring an ancient tradition and our natural resources, addressing human needs for shelter and comfort, and engaging in an inspiring, creative process with our clients. We invite you to participate fully in this process and trust that this comprehensive book, packed with valuable expertise and insight, will help guide and inspire you every step of the way.

Rick Bascom and Jeff Davis

chapter one

INTRODUCTION TO TIMBER FRAMING

"CONSIDER THE MONUMENTAL EVENT IN ARCHITECTURE WHEN THE WALL PARTED AND THE COLUMN BECAME."
- LOUIS KAHN

WHY A TIMBER FRAME HOME?

I was designing a large living room with cathedral ceilings, but something wasn't working. The huge space lacked human scale, definition and warmth. These are elements that are critical to how a room is positively experienced, and are especially important for a room where people gather for relaxation and leisure. A careful analysis of the options led me to the realization that timber framing would address those issues. By providing continuity, order and a way to define the different areas within the room, the timber frame gave the space a wonderful feeling and functionality that had been so elusive in other designs. While it was not ideal to arrive at this conclusion so far along in the design process, the discovery of a solution like this at a more advanced stage simply reminded me of how the design process

The timbers define a more intimate dining area without detracting from the room's grandeur.

is always iterative. For me, it also opened the door to the world of timber frame structures, with all their versatility, simplicity, tradition and beauty.

There are many reasons to choose timber framing over other methods of construction. Timber frames create a sense of strength, stability, longevity, warmth and comfort. They appeal in a nearly primitive way, addressing a subconscious desire to understand and trust the structure that shelters us from nature. Making use of the ancient craft of timber framing can also appeal to a more conscious sense of tradition. A timber frame home is an opportunity to uniquely express some of your personal values and aesthetics. For example, you might want to reuse handsome and aged reclaimed timbers, full of character from a previous life in another building. You might be attracted to an open floor plan or inspirational cathedral spaces. You might even be drawn on a more practical level to the higher-performance construction that the timber frame home offers.

Timber frames greatly enhance many aspects of a home's design. The visible timbers define and order spaces, provide variation in the ceiling planes and can change the scale of spaces both large and small depending on the frequency and size of timbers used. Timbers also have characteristics unlike man-made materials; variations in texture, color, size and even small defects make each piece unique. The use of wood imparts a sense of warmth to an indoor environment—it just feels good to be around it.

Whether your main reasons for choosing to timber frame are technical, aesthetic or otherwise, timber frames offer elegant, simple solutions to many of the challenges that exist in designing a home, especially those issues that are concerned with quality—not quantity—of space.

A BRIEF HISTORY

Some people wonder about the differences between post-and-beam and timber framing. Post-and-beam construction is a framework based solely on vertical and horizontal pieces: the posts and the beams. Post-and-beam structures are constructed in layers, with each floor built independently from the others. Typically one piece is erected at a time and simple or no joinery is used. Timber frame construction consists of posts that span unbroken from sill to eave or ridge and are erected in walls and bents. (Bents run perpendicular to the ridge of the roof, while walls run parallel to the ridge). Historically, timber frames were erected using pike poles and derricks and required a significant amount of manual labor. Timber framing also involves the craft of using wooden pegs and joinery to assemble the skeleton of the building. Joinery used to assemble frames in the 18th century and onward exhibit the use of metal, but the original and more traditional joinery was made entirely of wood.

Timber framing evolved concurrently in many cultures that had access to viable timber for harvesting. Some of the earliest examples can be found in Egypt as early as 2000 B.C., but historically timber framing was much more prevalent in Japan and Europe than in Africa. European timber framing is at the root of the American tradition, starting as simple structures with posts buried in the ground and then developing toward its pinnacle in the late Middle Ages. American architectural history saw the near complete abandonment of the craft in the 19th century but has fortunately been witnessing a grand resurgence of it in the past three decades. You may even want to investigate what was done traditionally in your region, and consider reflecting elements of that history in your own home.

EARLY NORTH AMERICAN FORMS AND THE SALTBOX HOUSE
EARLY NORTH AMERICAN TIMBER FRAME HOMES OFTEN BEGAN AS SINGLE ROOMS THAT WERE EITHER ONE OR ONE-AND-A-HALF STORIES HIGH WITH A FIREPLACE AT ONE GABLED END. THESE FRAMES WERE EASILY EXPANDED BY ANOTHER BAY ON THE OTHER SIDE OF THE CHIMNEY, WHICH GAVE THE INHABITANTS TWO ROOMS WITH A CENTRAL HEAT SOURCE. THE LATER ADDITION OF AN ATTACHED SHED ALONG AN EAVE WALL WAS SOMETIMES USED TO EXTEND THE FIRST FLOOR AREA, RESULTING IN THE DISTINCTIVE AND FAIRLY PREVALENT NEW ENGLAND SALTBOX HOUSE. THE SALTBOX HOUSE TYPICALLY HAS ONE EAVE WALL THAT IS ONE STORY HIGH AND AN OPPOSITE EAVE WALL THAT IS ONE-AND-A-HALF OR TWO STORIES HIGH. BOTH ROOF PLANES HAVE THE SAME PITCH SO THAT THE RIDGE OF THE ROOF IS OFF-CENTER BETWEEN THE EAVE WALLS.

Most early American timber frame homes were made from white oak, a wood familiar to early European settlers and chosen for its strength and natural resistance to rot and insects. Pine became the wood of choice in the late 1800s because of its abundance. Wattle and daub, which is a twig or plank infill coated with clay or mud, was often used for the walls of timber frame structures, but this gave way to clapboard and interior plaster walls because New England weather tended to be more extreme than what had been experienced in Europe. So began the evolution of timber framing in North America, and it continues to evolve today. Whether construction involves the integration of modern materials and building systems, the introduction of machine fabrication, or even a strict adherence to the craft as it once was, evolution is ongoing simply because the environment in which timber frame construction is performed continues to change.

Hammer beam trusses accentuate the form of the classic brick fireplace and chimney.

A PRIMER

Timber framing has its own language: bents, purlins, mortise and tenon are, for example, a few words that will be new to most people. There are also several common frame and joinery types, and different species of wood to choose from depending on structural performance, regional availability and your aesthetic preferences. The following is a basic primer about this terminology and these fundamental considerations.

FRAME TERMINOLOGY

Let's start with some of the frame terminology that will be used throughout this book. Terminology for joinery will be discussed a little further on.

Bent – A bent is the name given to a wall that runs perpendicular to the roof ridge, while walls are those that run parallel to the roof ridge.

Eave – Eaves are the lower, horizontal edges of the roof that typically overhang the walls.

Eave Plate – The eave plate is where the roof joins the walls and support rafters.

Eave Posts – Eave posts are posts that are part of the eave walls and support the eave plates.

Eave Walls – Eave walls are the typically rectangular walls that the roof rafters rest upon.

Post – Posts are the vertical load-bearing timbers used in timber frames.

Purlin – Purlins are timbers used to build roof structures that run parallel to the eaves of the home. They are used between principal rafters to help carry the roof load.

Rafter – Running perpendicular to purlins and eaves, rafters are sloped timbers that are used to help carry the roof load. Rafters are used in pairs with one on each roof plane. Rafters can bear upon the opposing rafter end, on a ridge beam or on principal purlins.

Rafter Tie – Rafters form two sides of a triangle, the bottom of which is sometimes formed with a horizontal beam. A tie is a horizontal beam that is placed part of the way down the rafters, creating a smaller triangle. Ties are used to prevent the eave walls from spreading and to the make the rafters more rigid.

Principal Rafter – Principal rafters are the main rafter pairs of a bent.

Principal Purlin – Principal purlins are two purlins on either side of a roof between the eave plates and the ridge. Where principal purlins are used, the rafters typically bear on the purlins and then continue on to connect with and bear on the end of the opposing rafter. Principal purlins are supported by internal posts.

Ridge – The ridge consists of a beam or beams that run the length of the roof, forming the roof's highest point. The ridge is at the top of the triangle formed by trusses or where the rafter pairs meet.

Truss – Trusses are structures consisting of vertical, horizontal and diagonal timbers that form triangular units. There are many well-known truss forms used in timber framing such as the king post or queen post truss, which will be explained further in the section about *Types of Frames*.

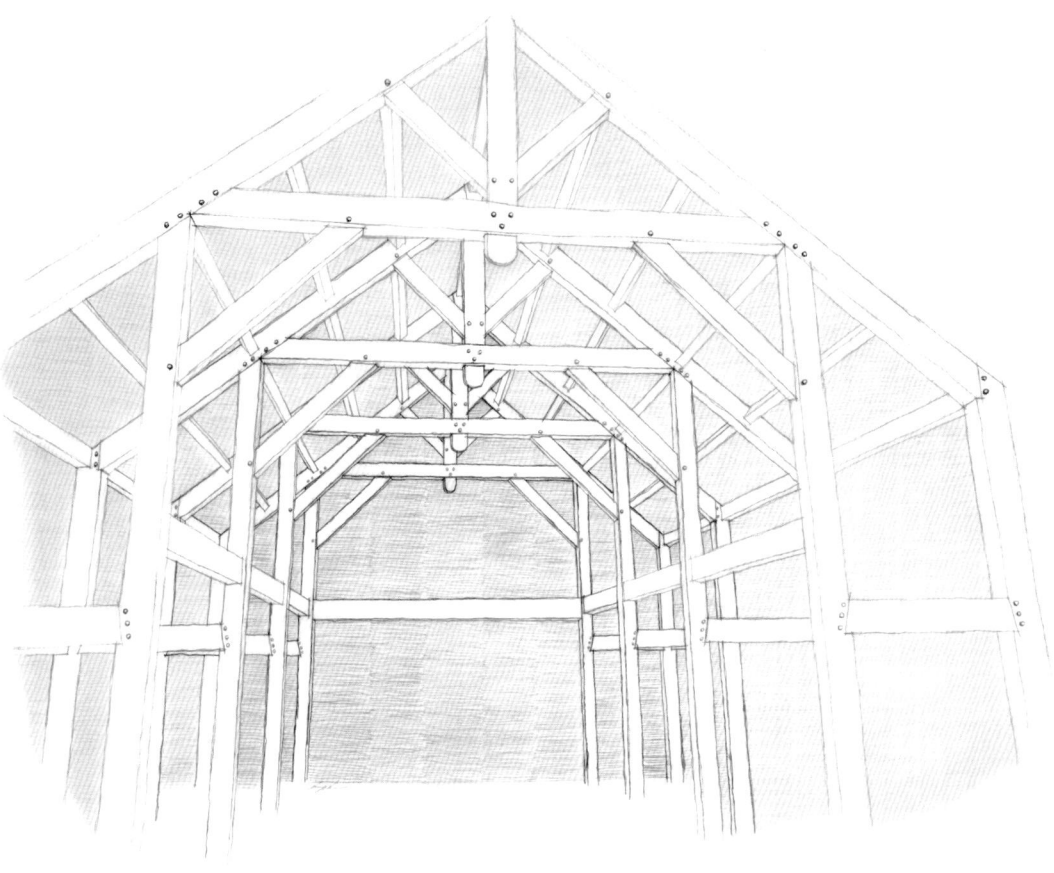

TYPES OF FRAMES

Most timber frame types built in North America today are either common rafter frames, or common purlin frames (also sometimes called bent frames) and combinations of both types exist. Although not all types of frames will fall into these two basic categories, they are by far the most common.

The following section discusses several traditional styles of these two common frame types but it is impossible to name every variety here, since timber frame forms are limited only by the imagination and budget.

HISTORICALLY, AISLED FRAMES WERE THE MOST PREVALENT EUROPEAN TIMBER FRAME TYPES IN NORTH AMERICA. THE TERM "AISLED FRAME" IS AN UMBRELLA TERM FOR ALL FRAME TYPES (INCLUDING BOTH COMMON RAFTER AND COMMON PURLIN TYPES) THAT CREATE AISLES IN THE FLOOR PLAN. PRINCIPAL PURLIN AND QUEEN POST BENT FRAMES ARE TWO EXAMPLES. AISLED FRAMES ARE OFTEN USED WHEN A FRAME IS REQUIRED FOR A SPACE THAT IS WIDER THAN THE DISTANCE THAT MOST RAFTERS AND TRUSSES CAN SPAN. FOR THESE WIDER STRUCTURES, RAFTERS AND TRUSSES GET EXTRA SUPPORT IN THE FORM OF INTERNAL POSTS AT APPROPRIATE INTERVALS. THESE INTERNAL POSTS CREATE AISLES ALONG THE EAVE WALLS OF THE FRAME, WITH A CENTRAL BAY REFERRED TO AS THE NAVE.

Common Rafter Frames

A common rafter frame is one in which the roof rafters span from the eave walls to a ridge beam, principal purlin or the opposing rafter. Common rafter frames are usually erected in sections of walls: principal purlin walls or ridge walls; and the eave walls. Perpendicular pieces are inserted as each wall is tilted into place. This process is followed by the installation of rafters, typically four feet apart, to complete the roof framing. When decking is applied to the roof of a common rafter frame, the boards must run perpendicular to the rafters.

Common rafter frames include collar tie frames, principal purlin frames and ridge beam frames, which are described below.

Collar Tie Frames

Collar tie frames have no internal posts and therefore create open spaces but can only be used for limited widths. Rafters span from the eave plate and bear upon the opposing rafter end at the ridge. The collar tie spans between rafter pairs in the lower third of the length of the rafters.

Principal Purlin Frames

Principal purlin frames use rafters that span from the eave plates, past a pair of main purlins, and then continue on to meet with and bear on the opposing rafter. Principal purlin frames can be considered a type of aisled frame because the internal posts that support the principal purlins form aisles that are parallel to the eave walls and a central bay or nave.

Ridge Beam Frames

Ridge beam frames use rafters that span from the eave plates to a central ridge beam or beams (depending on the length of the building). Ridge beam frames have posts located centrally to support the ridge beams and that must be considered in the floor plan layout.

INTRODUCTION TO TIMBER FRAMING

A hammer beam bent is being tilted into place by a crane. Purlins will then span the principal rafters to form the roof structure.

Common Purlin Frames

Common purlin frames consist of a series of purlins all bearing on the principal rafters of a bent. Common purlin frames are usually erected one bent at a time at uniform intervals. Perpendicular pieces are inserted as each bent is tilted into place. This process is followed by the installation of purlins to complete the roof framing.

When decking is applied to the roof of a common purlin frame, the boards must run perpendicular to the purlins.

Common purlin frames include hammer beam truss frames, king post bent frames and queen post bent frames, king post truss frames and queen post truss frames, and scissor beam truss frames, all of which are described below. Larger early American frame structures like barns, for example, tend to be versions of the king post truss and queen post truss frames.

King Post Bent Frames

King post bent frames make use of vertical timbers called king posts which span from floor to ridge. Diagonal braces from the king post to the rafters create the truss that helps support the roof load. The king posts must be taken into consideration when designing the floor plan.

King Post Truss Frames

A king post truss frame is similar to the king post bent, but with the addition of a tie that breaks the king post. The use of the tie transfers the loads to the eave posts and allows for a clear span by eliminating the king posts that would otherwise affect the floor plan.

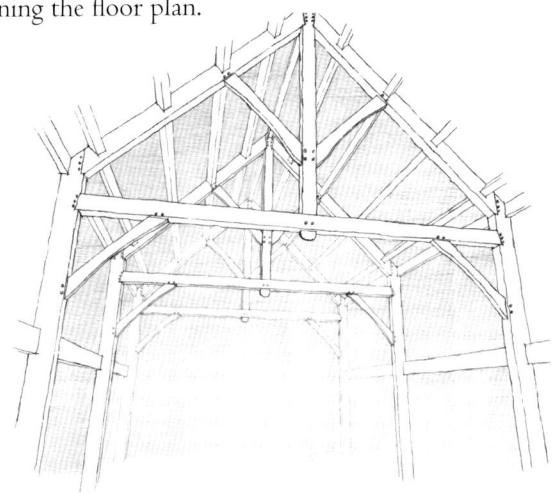

Queen Post Bent Frames

Queen post bent frames make use of vertical timbers called queen posts which span from floor to part-way along the length of the principal rafters. Similar to a principal purlin frame, the interior posts of the queen post bent frame will form aisles along the eave walls. At the top of each queen post, a tie beam is used to connect the post to the opposite queen post.

Queen Post Truss Frames

Queen post truss frames are similar to queen post bent frames, but with the addition of a tie that breaks the queen posts and transfers the loads to the eave posts. Like the king post truss frame, this allows for a clear span and there are no internal posts that will affect the floor plan.

Hammer Beam Truss Frames

Hammer beam truss frames are very similar to queen post truss frames, but two separate ties, called hammer beams, break the queen posts. Hammer beams exert great lateral pressure onto the eave posts, which must be taken into account in the design of the frame.

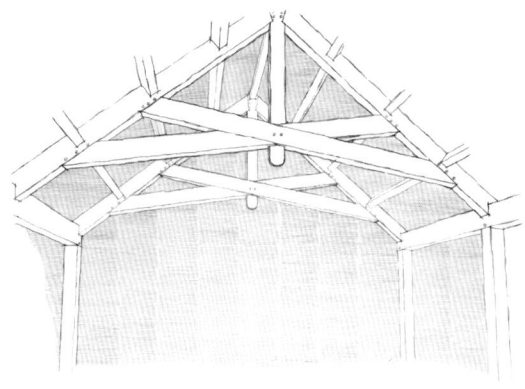

Scissor Truss Frames

Scissor truss frames use diagonal, crossing ties that span between opposing rafters. The struts of the truss provide a number of advantages; they transfer loads more efficiently to the eave posts; they reduce the thrust at the eave walls; and they allow for a greater span than just rafters alone.

Cruck Frames

Cruck frames are one of the oldest frame types known. The main timbers in a cruck frame are cut from trees that curve in a similar way. Each tree is then split down the middle so that the two halves form opposing arched timbers, lending a very organic look to the structure they create. A more recent innovation is to make cruck frames out of glue-laminated (or "glulam") timbers, which are made of smaller pieces of wood that are glue-laminated together. Although this may not sound attractive by description, glulam timbers can be beautiful and are very strong, lending themselves to architectural applications that would not be possible with plain, natural timbers.

THE PINE BOUGH TRADITION

IT'S AN ANCIENT TRADITION TO NAIL A PINE BOUGH TO THE PEAK OF THE FRAME WHEN A TIMBER FRAME HAS BEEN FULLY ERECTED. WHILE THE EXACT ORIGINS OR PURPOSE OF THIS TRADITION ARE UNKNOWN, WE CAN IMAGINE THAT IT WAS TO PAY RESPECT TO THE FOREST THAT THE TIMBERS ORIGINATED FROM, TO BRING LUCK TO THE HOME OR PERHAPS TO WARD OFF EVIL SPIRITS. THIS CUSTOM IS STILL OBSERVED IN MANY PLACES. IF YOU ARE INTERESTED IN DOING IT, NOTE THAT IT SHOULD BE PERFORMED BY THE OWNERS OF THE HOME.

A king post truss frame uses a tie to transfer the load to the eave posts, which leaves the floor plan open.

TYPES OF JOINERY

Timber framing joinery is a wonderful example of how beauty of form can be derived from a functional need. Conventional construction requires fasteners like nails and screws to hold pieces together. Timber framing typically relies on wooden joinery to hold timbers together, including mortise and tenon, tapered shoulder, half lap, dovetail, spline and scarf joints. The ends of timbers are carved and then wedded together, held fast by the mechanical properties of the joint or by wooden pegs.

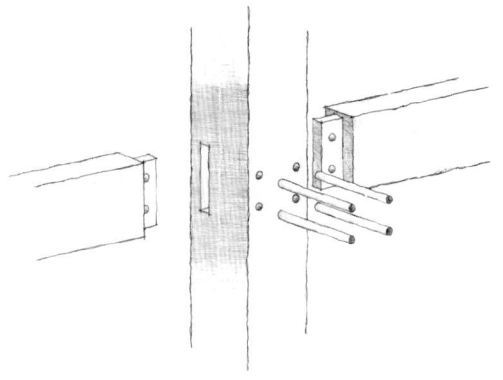

Mortise and Tenon Joints

These joints are the foundation of timber frame joinery. Mortise and tenon joints mechanically fasten two separate timbers and provide rigidity to the entire frame. Mortise and tenon joinery can be found in a wide range of connections in any type of timber frame.

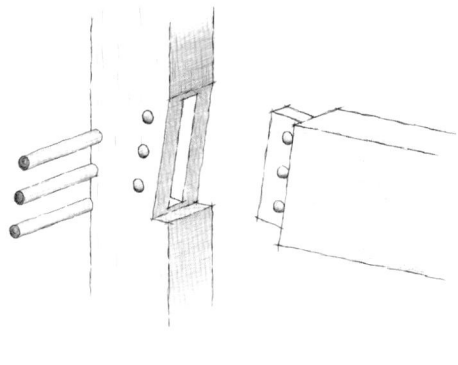

Tapered Shoulder Joints

The tapered shoulder joint (sometimes simply called a shouldered joint) is one of the most common bearing joints used to connect horizontal timbers and vertical posts. The shoulder provides a load-bearing surface that allows the horizontal timber to transfer its load to the post. The mortise and tenon then hold the joint together.

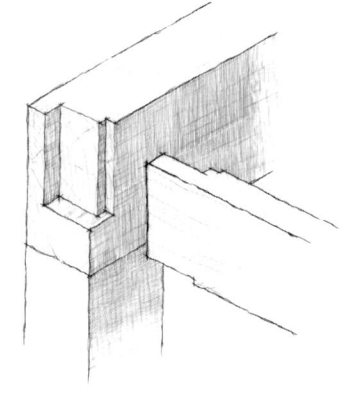

Half lap Joints

Half lap joints do not make use of a mortise and tenon but rely on removing equal and opposing amounts of wood from the two connecting timbers. The half lap can be secured with a peg or rely solely on the mechanical properties of the form of the joint.

Dovetail Joints

Dovetail joints have several applications in timber frames and are sometimes used, for example, to attach ties to rafters or posts. The dovetail joint relies on the mechanical properties of the interlocking tapered wood pieces.

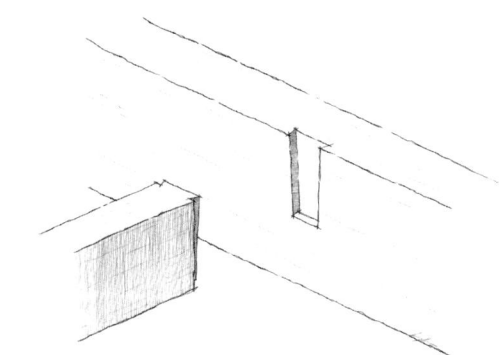

Scarf Joints

There are several types of scarf joints. Scarf joints are similar to half lap joints except that they are cut at shallow angles to the timber grain. A scarf joint relies on opposing wedges or a rectangular wooden peg, called a key, to hold the joint fast.

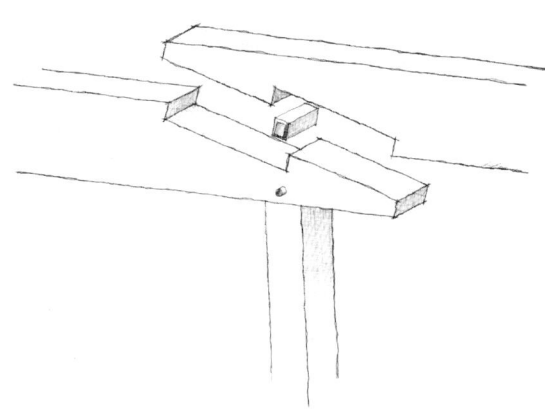

Spline Joints

Spline joints can be concealed or exposed depending on your aesthetic preference. A spline functions similarly to a tenon in that it is fitted into the matching mortises of two timbers. Spline joints are typically made from hardwoods and can be made of a wood that complements the color of the main timber wood, such as a contrasting tone.

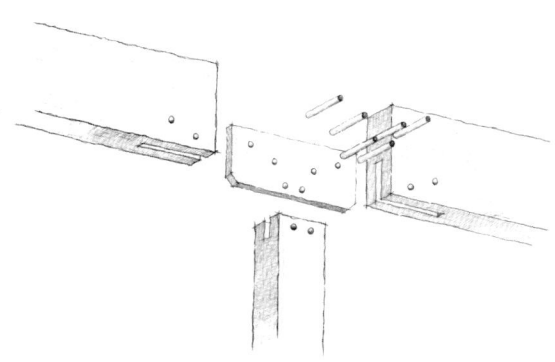

USING RECLAIMED TIMBER

IT IS POSSIBLE TO EXTEND THE LIFE OF TREES THAT WERE HARVESTED LONG AGO AND USED IN BUILDINGS. ASIDE FROM BEING AN ENVIRONMENTALLY FRIENDLY OPTION, RECLAIMED TIMBERS LEND UNIQUE CHARACTER TO INTERIORS.

Structural insulated panels are being tilted onto the frame. SIPs provide an unbroken thermal barrier to the elements.

TYPES OF WOOD

Modern transportation allows for wood from anywhere in the world to be used in timber frame construction. Gone are the days that a nearby river was necessary to transport harvested timber. Your choice of wood will be based on factors that include cost, sustainability, desired aesthetic and performance. Each will have to be weighed carefully alongside each other and each choice will have its own impact on the design process.

Oak, pine, hemlock and Douglas fir tend to be staple choices for today's frames and each comes with a set of mechanical and aesthetic characteristics. Oak is heavy, has a historic resonance and an envied grain pattern, but it is not necessarily the strongest wood once its weight is accounted for, and can be more expensive. Pine and hemlock have regional appeal almost throughout North America, are both readily available and fast-growing, and both lend themselves to a rustic aesthetic. However, they are some of the weaker woods. Finally, Douglas fir is strong softwood, and has a clean grain and character, but because it is most readily available on the west coast of North America, transportation costs could be prohibitive, depending on where you live.

TIMBER FRAMES AND INSULATED PANELS

While the timbers address the most important concerns about structure and form for a timber frame home, there is still the question of what will be used to enclose the spaces and structures created by the timbers. At Davis Frame, we have come to believe that the use of structural insulated panels (SIPs, an acronym typically pronounced as a word: "sips") in home construction beats out more traditional methods for creating walls and ceilings. While the timber frame addresses the needs of the spirit, the insulated panels address the practical need for efficient high-quality construction. We think of it as a harmony between tradition and innovation.

SIPs offer many common advantages over other wall and ceiling materials, which include reduced drafts as well as better thermal and sound insulation. However, the largest single benefit is energy efficiency because they provide a continuous thermal barrier between the interior conditioned spaces and the exterior environment. This continuous thermal barrier is possible due to the way SIPs are constructed. They consist of a core of rigid insulation with sheathing on one or both sides. The types and varieties of the sheathing on the panels are numerous, as are the types and thicknesses of the rigid insulation at the core. Different types of insulated panels will be discussed in Chapter 3 in the section about *Building Shells*.

TYPES OF CONSTRUCTION

Most typical homes today are constructed using platform framing. This type of framing originated from and is very similar to balloon framing. Both types of framing use small uniform lumber in increments placed about 16 inches (40 centimeters) or 24 inches (60 centimeters) apart.

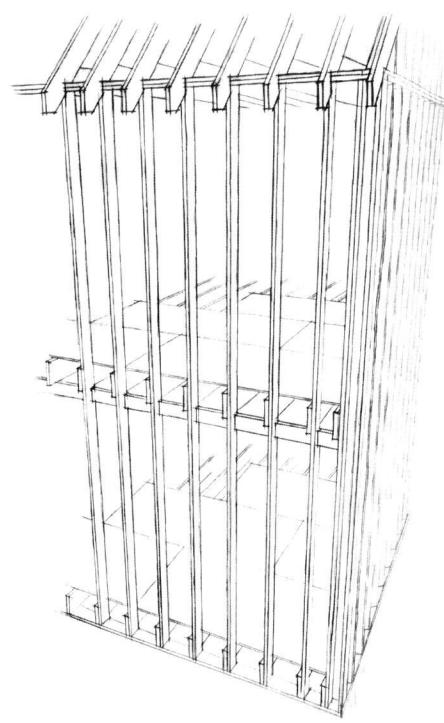

Balloon framing consists of wall studs that span from foundation to eave with a floor platform suspended within the frame.

BALLOON FRAMING

Balloon framing consists of wall studs that span from foundation to eave with a floor platform suspended within the frame. Balloon framing became prevalent in the United States in the 1830s in Chicago, when lumber, water-powered sawmills and machine-made nails were plentiful. The greatest appeal of balloon framing was that it was an easier craft to master, making it a very popular construction method in rapidly growing boomtowns. This was also why the use of timber framing almost completely ceased around this time.

PLATFORM FRAMING

Platform framing is similar to balloon framing but consists of wall studs spanning from the first floor to the underside of the second floor. A platform is then erected at the top and the second story is built from there. Successive stories are built in a similar manner, typically to a maximum of four. Platform framing evolved from balloon framing primarily because in the context of boomtown building, long lumber pieces were not as readily available and the platforms made construction more efficient from a labor perspective.

PLATFORM FRAMING COMPARED TO TIMBER FRAMING

The downsides of site-built platform framing lie in the costs and inefficiencies associated with on-site construction in general. Off-site fabrication used for timber framing is performed in controlled environments that are not exposed to the elements and can deliver significant benefits both in terms of efficiency and quality. Because timber frames and SIPs are both manufactured off-site and then assembled in the field, it can greatly reduce the on-site time required for completing the shell of the home.

ON-SITE VERSUS OFF-SITE WORK
THE PREPARATION OF BUILDING MATERIALS AT THE CONSTRUCTION SITE CAN BE INEFFICIENT AND COSTLY AS WELL AS MORE WASTEFUL OF RESOURCES BECAUSE THE MAJORITY OF MATERIALS MUST BE SHIPPED THERE AND WASTE MATERIALS MUST BE SUBSEQUENTLY SHIPPED AWAY, OFTEN AT A PREMIUM COST. WHEN PREPARED IN A SHOP OR PLANT, ONLY THE REQUIRED MATERIALS WILL BE SENT TO THE CONSTRUCTION SITE, AND SCRAP MATERIAL CAN BE EASILY REUSED OR RECYCLED AT THE POINT OF MANUFACTURE.

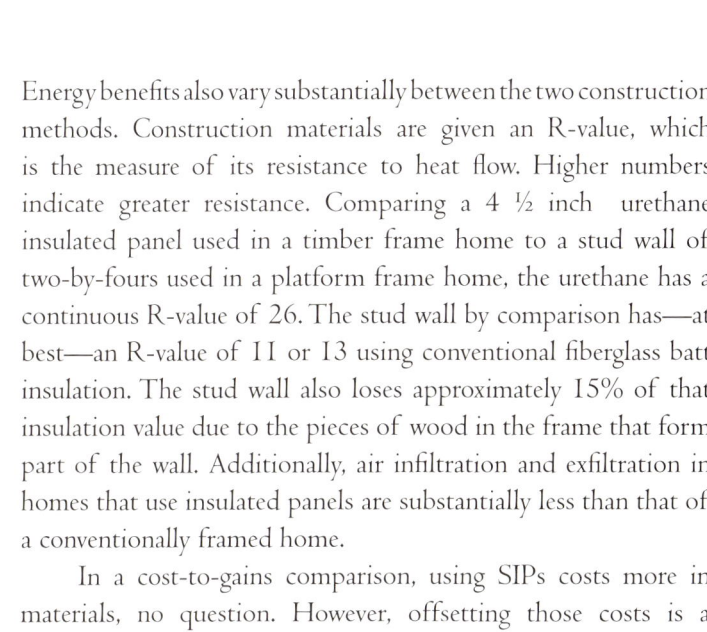

Although a natural looking tung oil finish is very common, a dark stain against light walls can create a striking effect.

Energy benefits also vary substantially between the two construction methods. Construction materials are given an R-value, which is the measure of its resistance to heat flow. Higher numbers indicate greater resistance. Comparing a 4 ½ inch urethane insulated panel used in a timber frame home to a stud wall of two-by-fours used in a platform frame home, the urethane has a continuous R-value of 26. The stud wall by comparison has—at best—an R-value of 11 or 13 using conventional fiberglass batt insulation. The stud wall also loses approximately 15% of that insulation value due to the pieces of wood in the frame that form part of the wall. Additionally, air infiltration and exfiltration in homes that use insulated panels are substantially less than that of a conventionally framed home.

In a cost-to-gains comparison, using SIPs costs more in materials, no question. However, offsetting those costs is a significant return on investment that will be realized in the reduced energy costs of the building, both immediately—because you can purchase smaller and more energy-efficient heating systems—and over the long-term, because of reduced energy costs over the life of the building. Costs are also offset somewhat in that the use of SIPs means that the shell of the building is assembled much faster to protect the structure from the elements sooner.

SIPS AND ENERGY STAR RATINGS

THE USE OF SIPS HAS BECOME WIDELY RECOGNIZED AS FAR SUPERIOR TO OTHER METHODS OF SEALING AND INSULATING BUILDINGS. FOR EXAMPLE, IF YOU USE SIPS IN THE CONSTRUCTION OF YOUR HOME AND THEN APPLY FOR AN ENERGY STAR RATING, YOU ARE NOW AUTOMATICALLY EXEMPT FROM THE BLOWER DOOR TEST, WHICH MEASURES AIR INFILTRATION TO A BUILDING AS A WHOLE, BECAUSE THE PERFORMANCE OF SIPS HOMES IN THIS REGARD IS SO UNIFORMLY AND CONSISTENTLY HIGH. THE TEST IS PERFORMED BY SEALING A FAN INTO AN EXTERIOR DOOR OPENING, AND THEN MEASURING THE AMOUNT OF AIR THE FAN IS ABLE TO PULL INTO THE SEALED HOME.

CHOOSING TIMBER FRAME AND JOINERY TYPES

You're reading this book because you're interested in building a timber frame house, so you may be wondering which type of frame to choose. Perhaps you will be relieved to know that unless you have a very specific type and preference already in mind, you do not have to choose yet—or even for quite some time.

There are some factors that will determine what your options are. For wide spaces, you will be limited to frame types that support very large spans, such as principal purlin, king post bent or queen post bent frames. Smaller spaces can be created with narrower frame types such as ridge beam and collar tie frames. For medium-width spaces, king post truss, queen post truss and scissor truss frames are the most commonly used, because they allow for open floor plans. It is common for more than one frame type to be used in a single home.

If you have a preferred frame type, then this can of course be accommodated in the design by basing other decisions about the home on the use of that frame type. Be sure to communicate these preferences to your designer or architect as early as possible, but know that final decisions about types of frames are seriously discussed much farther down the road in the schematic design phase, and then are formally defined in the design development phase, both of which are outlined in Chapter 4.

Similarly, joinery decisions will be made based on a number of factors including: how all of the pieces will be assembled at the site; the type or types of frame being used; the amount of compression or tension on the joints; the effects of wind, snow and gravity on the frame; and of course your aesthetic preferences. Some joinery types are more intricate than others. I prefer simple clean joinery that is straightforward in its function and purpose. As with framing decisions, joinery choices are made during later stages of design development.

Finally there are finishing-touch choices such as whether to leave timbers rough or have them sanded and rounded at the edges, and whether to stain or not to stain. Frames can be left untreated, but if so the grain of the wood will more readily accept dirt and dust making them more difficult to clean. I recommend finishing timbers with tung oil, which soaks into the pores of the wood and seals the surface while retaining a natural look. A second coat of tung oil can be applied at the site once the frame has been assembled and cleaned and the interiors finished. If you like, stains can be mixed into the tung oil for the second coat. Surface sealants like a varnish can crack and peel and are therefore not recommended.

This great room was too wide for a collar tie frame, but we were able to use a king post truss to allow for an open floor plan.

chapter two

GETTING STARTED

"...DESIGN IS NEITHER AN INTELLECTUAL NOR A MATERIAL AFFAIR, BUT SIMPLY AN INTEGRAL PART OF THE STUFF OF LIFE..."
- **WALTER GROPIUS**

DOING THE GROUNDWORK: EXPLORATION

To build a home that you will love, and before the formal design process can begin in earnest, you must start with a vision. Developing your vision takes time, and it doesn't mean that you can't adjust and revise your ideas once you've started the formal design phase. However, beginning with a fairly comprehensive sense of what you want will support the whole design process and allow things to run more smoothly; and most importantly, it will help ensure that the home you build is right for you.

While a home obviously serves many functional purposes, it can also meet a wide range of unique personal, aesthetic, emotional and other requirements that are more difficult to describe. Despite the difficulty you may have in articulating and defining some of these less concrete requirements, it does help if your vision for your home takes as many of these aspects as possible into consideration. Let's

look at a few things you can do that can help you identify and communicate what and how you want your home to be.

It can help to think of the vision-building process as an exploration or a new adventure. This attitude can even turn walking down familiar streets into something fresh, because you will be focusing on things that you probably didn't have much reason to notice before. For example, at a very basic level, you may not be someone who normally spends a lot of time thinking about doors; but if you want a customized house, you will certainly want to think about doors for a little while. You might even suddenly find yourself noticing doors; doors you like, and doors you don't like. When you do see one you like, take the time to reflect on what is it about the door that has caught your eye. It may be attractive to you because it is different in some way from other doors you have seen before. Maybe it is because of the color or carving of the wood, or because of the materials or even the hardware used. If you use a similar approach to thinking about other aspects of a house, such as the walls, ceilings, lighting and windows, you will find yourself becoming more attuned to your preferences and inclinations, and you will start noticing sources of inspiration all around you.

Aside from things like doors and windows, there are other, less tangible things to identify. You may find yourself asking the question, "How do I describe the spaces and feelings I want my home to have?" The answer is, "Describe them by every means possible." Bring a notebook with you for jotting down ideas. You might consider carrying a digital camera to take pictures of things that catch your eye—a window, a room, a nook, an arch.

Look at photographs of favorite places, gatherings and vacations. When looking at photos from gatherings in particular, identify the spaces that the people in the photograph have gravitated towards. Where can the groups of people be found—in the living room, on an outdoor deck or in the kitchen? Cut out pictures from magazines or mark pages in books. Use a marker to circle things, use sticky notes—anything that can be used to record your thoughts and impressions. Write descriptions of what images represent to you, and note positive, negative and neutral aspects. Describe the feelings that images evoke using whatever terms are most appropriate for you, like: too tall, too dark, cozy, warm, inviting, or airy.

You can also think about spaces that evoke for you the feelings that you want your home to have—even examples of places that feel wrong to you can be useful. Think of public spaces like parks or libraries, or private spaces, like the homes of friends and family members. For example, you might notice a small park with a courtyard that provides a perfect intimate outdoor space for one or two people to have lunch. This space could become the inspiration to design a cozy intimate exterior space for your new home, like a balcony or small patio adjacent to a bedroom. As another example, you might realize that your local public library provides no intimate locations for a person to sit and read; perhaps the chairs are in the middle of a large room under a high ceiling with poor acoustics and glaring western light. Being clear about your dislike of this loud and harshly lit space can help you identify what kinds of spaces you do want to create in your home instead.

USE A NOTEBOOK OR START A BINDER THAT YOU CAN DEDICATE TO YOUR HOUSE-DESIGN IDEAS, PHOTOS, CLIPPINGS, NOTES AND LISTS. THERE ARE A NUMBER OF DIFFERENT EXERCISES LATER IN THE BOOK WHICH HAVE BEEN DESIGNED TO HELP YOU VISUALIZE AND ARTICULATE WHAT YOU WANT IN YOUR DREAM HOME. YOU CAN INCLUDE THESE IN YOUR NOTEBOOK. THE MORE CLEARLY YOU CAN SHOW OR COMMUNICATE WHAT YOU WANT TO THE DESIGNERS OF YOUR HOME, THE HAPPIER YOU'LL BE WITH THE FINAL RESULT.

While exploring, assess the qualities as well as the quantity of different spaces. The latter is the easier of the two to define. If you don't have a tape measure with you to determine the exact size of a room you like, you can estimate dimensions by counting floor or ceiling tiles. Floor tiles made of vinyl, ceramic or similar materials are often about 1 foot (30 centimeters) squared, while ceiling tiles are often 2 feet (60 centimeters) squared or 2 feet by 4 feet (60 by 120 centimeters). Ceiling height can have a significant impact on the perceived size of a room so you should estimate or measure this dimension as well.

Average interior doors are close to 7 feet (2 meters) high and typical homes have 8-foot (2.5-meter) ceilings, leaving only a foot (30 centimeters) or so of wall space above the head of the door. Note too if the doorways are arched or square cased, since these also have an impact on the feeling of a room.

Defining the quality of a space is a more difficult task, because how a place feels depends so much on individual sensory experiences. While architecture communicates primarily in the realm of the visual (creating structures and spaces one can see and interpret), sight is not necessarily every individual's

predominant sense. Some people are much more sensitive, for example, to touch or sound.

As an example, consider how a room that is small in scale and separate from the louder areas of a home may appeal to someone whose sense of hearing is their predominant sense. Alternatively, the brightness or color of a room may have a greater effect on someone whose sense-predominance is visual. If more than one individual will be living in the home, identify what sensitivities are the most important to address and see if you can correlate these things to different types of spaces.

While looking at a photograph of a bedroom in a magazine, for example, it might be the soft light coming through windows that attracts your attention and makes the space seem warm and inviting. If the walls of the room are soft in color, the ceiling is of moderate height, and the room is just large enough for the bed, dressers and a single chair by the windows, it might feel quiet and cozy. If you can understand what appeals to your senses and to your idea of what is appropriate for different spaces in the house, it will help you express what you want your home to be like.

IF YOU HAVE COMPILED A COLLECTION OF IMAGES AND IDEAS, GO THROUGH IT AT THIS STAGE LOOKING FOR UNIVERSAL SIMILARITIES AND DIFFERENCES, MAKING NOTE OF THOSE QUALITIES. KEEPING YOUR VISION IN MIND, SORT AND LABEL THE IMAGES YOU HAVE COLLECTED FOR RELEVANCY. IF MANY ITEMS ARE SIMILAR IN NATURE, TRY REDUCING THEM TO SINGULAR STATEMENTS. IF, FOR EXAMPLE, MANY OF THE PHOTOGRAPHS PERTAIN TO WINDOWS AND THE QUALITY AND QUANTITY OF LIGHT IN DIFFERENT ROOMS, YOU MIGHT COME UP WITH A STATEMENT LIKE: "WE WANT AS MANY ROOMS AS POSSIBLE TO HAVE SOUTHERN EXPOSURE BECAUSE NATURAL LIGHT IS VERY IMPORTANT TO US." THIS EXERCISE WILL HELP YOU SHARE YOUR IDEAS MORE EASILY WITH YOUR DESIGNER OR ARCHITECT AND HELP REFINE YOUR VISION.

Interior Design by Deborah Timmermann, ASID, Nantucket, MA

DEFINING YOUR VISION

Now that you've spent some time exploring what you like and dislike, the time has come to state it clearly and with a specific purpose. Your vision statement will provide the guiding principles to which you will turn when faced with difficult design decisions. Your vision will also stand as the basic benchmark against which the design will be held throughout the design process.

Can the vision change as the design process evolves? Absolutely—but it doesn't have to. The most important thing is to start out with a vision and continually revisit it. Think of using your vision statement the way you'd use a map on a journey to a specific destination. In this case, the destination is your custom home. Before any journey, it makes sense to have a route plotted before you begin, but as circumstances arise you may discover a need to amend the way you get there or reconsider your exact destination.

You can create your vision statement in a range of forms: from a single decisive sentence that expresses your purpose to a long list of more loosely defined desires and requirements.

Here are two fictional examples of rough visioning statements:

---

1. "I want my home to reflect my sense of stewardship for the environment; from every material and finish choice to how it sits on the site."

---

2. "Our home should be comfortable for our family of five. It should also address the future needs of our family as it grows. We want the living spaces to face the meadow, just like my parents' home in Vermont did. We want our house to look like it has been there for generations. We want our children's rooms on the first floor. We want the ability to add on a future sunroom on the south-eastern side of the house. We are indifferent about an attached garage."

While both examples differ in form and content, they both establish clear guiding principles to follow. They address desires that are both tangible and less tangible.

IMPORTANT CONSIDERATIONS

Once you have identified what you like and have defined a basic vision for the home, there are a few more things to be conscious of before you dive into the design process. These other variables include, again, both the tangible and less tangible. I have divided these here into "fundamental" and "elemental" considerations.

FUNDAMENTAL CONSIDERATIONS

The following are some of the more tangible variables that should be considered when you're planning your custom home: the individual inhabitants, your perceived future of the building and your budget.

INDIVIDUAL INHABITANTS

Homes function as an extension of their inhabitants. In a custom home, the design can reflect more deeply their individual needs and desires. Is the home for a single person, a couple, a family? Are there accessibility issues, such as requirements for wheelchair access, single-story living or the ease of use of fixtures and doors? Is anyone elderly? If possible, input from everyone who will live in the home should be sought and considered.

FUTURE OF THE BUILDING

What will your home become over time? There are no certainties in this regard of course, but you can consider probabilities. Timber frame houses are very solid and can last for many generations. Looking at the shorter term, perhaps spaces designed for children require the flexibility to be reconfigured as the children grow up. Another option is to consider where future additions might go as you are completing the design process, in order to balance the demands of present budget and future growth.

BUDGET

You'll have to manage your budget throughout the design process. A great early step is to talk to builders in the area where the home will be built. Ask about approximate average costs of construction per square foot and find out what that cost consists of and how it is calculated. For example, does the quoted amount reflect a home with inexpensive windows, inexpensive finishes and a simple rectangular floor plan, or something more elaborate?

In all cases, always allow for contingencies. The recommended buffer is usually about 10 percent of estimated costs for the house, to cover things that you may change or that

require greater work than originally estimated.

Another way to support yourself in managing your budget is to keep in mind, when describing the characteristics of your future home, the difference between wants and needs. Both are important. The wants will provide you with some flexibility for the design, and in many cases will make it possible for the needs to come to fruition. For example, the "want" of four bedrooms may allow the "need" for more square footage for the kitchen and family room, by simply combining the fourth bedroom with the home office—letting this one room do double-duty. Needs define the key priorities for the design; wants identify which items must be reconsidered if the budget is exceeded at any point in the design process.

ELEMENTAL CONSIDERATIONS

The following are some of the less tangible, the "elemental" variables that should be considered in all aspects of design: how light is used, how views are framed and how a sense of permanence is created. In my career as an architect, I have come to believe that these three things underlie the most inspired designs. Incorporating them thoughtfully is one of the main goals and guiding principles of my work.

Light or the absence of it is a key element in the perception of a space.

LIGHT

"ALL MATERIAL IN NATURE, THE MOUNTAINS AND THE STREAMS AND THE AIR AND WE, ARE MADE OF LIGHT WHICH HAS BEEN SPENT, AND THIS CRUMPLED MASS CALLED MATERIAL CASTS A SHADOW, AND THE SHADOW BELONGS TO LIGHT." – LOUIS KAHN

Light makes visible the spaces that we inhabit. It has characteristics similar to those of tangible materials—including texture, depth and color—which are as equally important to consider as the things that it will illuminate. Light or the absence of it is a key element in the perception of a space.

For these reasons, the design of a home must take into consideration the impact of light on all of the spaces it contains. The intended use for a space and the times of day it is likely to be occupied will help determine how much natural and how much artificial light is appropriate. Light can be used to draw the eye or focus attention on specific areas of a room, or it can play a supporting role, gently illuminating a space with an even light.

The sun is the most important source of light to consider in your design. We are creatures of nature and benefit enormously from the sun's warmth and light. The light of the sun has characteristics that change throughout the day and throughout the seasons, making it a steady and dynamic contributor to the design process. Sunlight is the first and most efficient use of solar energy, and designing a space to use daylight offers the biggest return on investment because it's free!

Views

"Always design a thing by considering it in its next larger context—a chair in a room, a room in a house, a house in an environment, an environment in a city plan." - Eliel Saarinen

Views refer to how the house is revealed to you as you approach it, as you move through it, and of course, what you see when you look out from it. While this will be discussed more fully later in this book, it is worth introducing the general concept now because it is so important to the overall experience of a home.

As an example, consider a home from the first time it is seen from the road or driveway. Does the home sit proudly atop the land or does it nestle into its surroundings? Is the entrance visible and is it inviting or austere? When you enter the home, do you enter into a grand foyer, a great room, or a modest hall with only partial views that invite your exploration of the adjoining spaces? Think of the journey through your home from the first moment it is seen, right up to the discovery of its most intimate alcoves. Spaces in a home might be formal, inspired by basic symmetry, traditional designs and historical configurations; or spaces might feel more organic. Think of the views that will be created from each space into the next, from room to room, from inside to outside and from outside to inside.

Views from inside to outside might capture panoramic expanses of countryside, mountains or ocean; or they might peek into an intimate side yard garden with only a bench, a few flowering fruit trees and a small stone terrace. Realize too that views from the inside to the outside give us a spatial sense of being sheltered.

Another thing to consider is that all views, from the intimate to the grand, whether indoor or outdoor, are partly defined by the spaces in between, in the transition spaces from one to the other, from private to public. Transitional spaces create a frame for the view that lies beyond, so give some thought to hallways, passages and doors.

Timber frame homes exude a sense of structural security and permanence.

Permanence

"Architecture aims at Eternity." - Christopher Wren

We care for the things that reflect the values we cherish most. In terms of architecture, the places that garner the attention, money and effort required to preserve them include historic districts, homes of famous historical figures, churches, town halls, and works by notable architects and designers. In North America we measure the ages of our treasured historic buildings in hundreds of years, but the ages of globally historic structures are often measured in thousands of years. What is the underlying characteristic that inspires people to fight for the survival of these buildings? I believe it is a sense of the building's permanence that is both relative and relevant to our society.

Architect and author Sarah Susanka referred to the majority of today's conventional homes in her book *The Not So Big House: A Blueprint for the Way We Really Live* (Taunton Press, 1998), describing them as "…a collection of massive storage containers for people." I believe this is true. Few homes built today are anything more than a plan brought into three dimensions. Little thought is given to how spaces in the home will be experienced, or to whether people will be able to perceive how the shelter above and around them stands.

In recent North American history, a real shift has occurred away from craft-based construction in home-building. This is not to say that there is no longer any craft involved in the design or construction of homes, only that craft is very rarely part of the structure of our buildings and houses. However, if you look at historic buildings internationally, such as the Parthenon or the Temple of Athena Nike, it is easy to see that these were constructed using simple structural elements using the craft of column and lintel. Later architecture, including the great cathedrals of Europe like Westminster Abbey, was constructed with massive buttresses and more complex arches and heavy pillars. Timber frame homes, with their solid, exposed frames made of wood express a similar sense of structural security and permanence.

CREATING AN OUTLINE

The following steps will assist you in identifying what is most important, to whom and why. Get out your notebook and start writing down your thoughts and ideas as you go through this chapter.

GUIDING QUESTIONS

For every single room that you plan, ask yourself the following questions and consider both the fundamental and elemental considerations discussed earlier:

1. What are the important adjacencies or functional relationships to other rooms?

2. Where on the site should the room be?

3. What are the functional requirements—or how will the room be used and by whom?

4. What are the most important features of this room?

5. How should this room feel or look—what are the defining characteristics of this space?

6. What are the area requirements in terms of square feet or dimensional limits or minimums?

Using a mudroom as an example, here are some sample answers to give you an idea of what type of issues you might find yourself addressing, even if some of the answers repeat themes already addressed.

1. What are the important adjacencies? The mudroom needs to connect to the garage and the kitchen. We also would like it to access the back yard and we would use it as an informal entrance to the house.

2. Where on the site should the room be? Off of the garage and with access to the back yard.

3. How will the room be used and by whom? We'd like the children to have cubbies for their clothes and boots when they come in from the back yard. Their friends can use that entrance instead of the formal entrance. Also, we need it to be connected to the garage so that it's easy to bring groceries in.

4. What are the most important features of this room? The cubbies and a bench are very important, because we don't like clutter. We want to have a nice back yard and it would be great if parts of it could be seen from the mudroom.

5. How should this room feel or look? We'd like it bright, with light colors and a lot of glass if possible. A durable floor of tile may be useful.

6. What are the area requirements? We'd like six 2-foot (1/2-meter) cubbies because there are four family members; this leaves two for guests. This means at least one wall should be 12 feet (3.5 meters) wide.

Grab your house-design ideas notebook and go through these questions for everything you think is relevant in the following lists:

THE SITE

A home should look like it belongs in its surrounding environment. The home should embrace the strengths of the site, whether these are gentle grades or steep hills, panoramic views or small and intimate visual focal points. Anything you perceive as a weakness of the site is an opportunity to be creative and find a way to diminish or transform the deficiency through ingenuity in design.

Consider the following about your site:

- Topography
- Access
- Utilities
- Landscaping
- Garage
- Breezeway
- Entry
- Porch
- Deck
- Patio
- Materials

The topography of a site will help determine the placement of the home, driveways, paths, gardens, garages and utilities. Natural features such as rock outcroppings, steep hills, water and vegetation in conjunction with a consideration of prevailing winds and the effects of sunlight will help in determining the best orientation for the house on the lot as well as its placement in relation to these features. Examples include using evergreen trees to shelter the home from colder prevailing winter winds or using deciduous trees to allow warming light in winter and shade in the summer. Moving water may offer a focal point for the house and site design, as well as providing a pleasant sound buffer from other noises.

Site topography will also help determine where on the site utilities are best located. Wells for potable water, septic systems and fuel storage are not typically considered "design" features, but lack of planning in regards to their placement can lead to poor design overall. Consider how power, phone and cable will

GETTING STARTED | 53

This building site required a creative driveway—but it was worth the extra effort.

be brought to the site: if the lot has a view in the direction of the street, for example, you can plan it so that utility lines and poles don't mar the aesthetic of a beautiful meandering driveway.

Now, what about driveways? Does the lot allow for vehicle access with a minimal amount of grading? Does the drive lead directly to a garage? Do you want to have room for guest parking, and how many vehicles should be accounted for? Driveways do not have to take the shortest route possible and can be considered the first part of the journey into your home. At the end of the driveway, is the garage prominent, hidden, or quietly secondary to the home? Is the garage attached? Detached garages are sometimes desirable, depending on the climate. If the garage is separate from the home, consider breezeways, paths or trellised walks from the garage to the home. (We'll discuss garages in more detail in the next section about *Utility Spaces*.)

Where is the main entry for the house? Is a formal entrance required and is it approached by a path, or does the drive allow for people to be dropped off in front of it? Who uses this entrance and will it be used frequently? Is an informal entry into a mudroom, for example, more appropriate to your lifestyle, or are

IF INDOOR AIR QUALITY IS A PARTICULAR CONCERN, CONSIDER INSTALLING A CENTRAL VACUUM THAT WILL EXPEL ITS EXHAUST DIRECTLY OUTDOORS.

This home has a traditional entrance centered within the facade and sheltered by a classic farmer's porch.

both formal and less formal entries required? Formal entrances and living spaces have been losing favor in home design in the last few decades. Most people in North America today simply do not have a lifestyle that requires grand foyers and receiving parlors. From a guest perspective, formal entrances work well if guests will be parking nearby or if street access leads to the entrance. Inhabitants, however, will more often enter through side, back or garage entrances because these will be closer to where they typically park their vehicle. An efficient design might have the garage, outdoor entrance and main entrance to the house all open to the same room, like a mudroom or closed foyer, allowing for a single air-locked entrance which will consolidate the majority of outdoor pollutants such as dirt dust and pollen into one space.

Porches, decks and patios can extend your living space and blur the lines between outdoor and indoor. They create a transitional space because they are outdoors but are sheltered and attached to the home. Outside rooms can become thematically or physically part of the outdoors. These outside rooms can be either private or public spaces. Placed in view of neighbors or the street, these spaces become areas where social interactions can easily occur, like neighborly conversations and even impromptu gatherings. More intimate spaces require separation from public view and sometimes even from adjoining areas of the home. A small balcony off a master suite, for example, should have visual privacy from the street and neighbors, and possibly privacy from the yard of the home.

When thinking about porches, decks and patios, decide which indoor areas they will be connected to. Are they meant to be an alternative to the indoor space or will they be used as overflow space for entertaining? Will they be part of how people are welcomed into the home? Porches have to be carefully considered because their roofs can block light to the first story—but remember that a porch might be as simple as a gabled roof

SITE STEWARDSHIP

YOUR HOME AND SITE WILL BE BECOME SYMBIOTIC PARTNERS, SO CARE FOR YOUR SITE SHOULD BEGIN DURING THE CONSTRUCTION PROCESS. DISTURB AS LITTLE SITE AS POSSIBLE DURING CONSTRUCTION, PROTECT THE UNDISTURBED AREAS WITH EROSION CONTROL MEASURES AND IDENTIFY AND PROTECT TREES AND VEGETATION THAT YOU WISH TO PRESERVE. WHEN PLANTING GRASS AND VEGETATION; PLANT GRASSES THAT WILL REQUIRE LITTLE WATER AND CHEMICAL SUPPORT AS WELL AS CHOOSING NATIVE PLANTS ACCUSTOMED TO YOUR SPECIFIC SITE AND SOIL CONDITIONS, REDUCING OR NEGATING THE NEED FOR IRRIGATION.

above a doorway, as traditional as a wrap-around, or grand as a porte-cochere that will shelter a vehicle and its occupants when they arrive at your door. Screening also affects how a porch can or will be used.

Finally, your choice of materials will affect the visual relationship of the house to its surroundings. For example, if your house reflects a particular architectural style, will you choose the corresponding traditional materials or try something different? Natural cedar shingles weather quickly to a silver-grey, and if you add white trim it will suit a New England-style cape house. Perhaps for a home in the woods you'd prefer heavier and rougher materials in deep earthen tones. You can also choose materials that artfully contrast with the surrounding environment and highlight features like the entries and windows. Whatever materials you choose, look for durability and maintainability, so that they will last a long time—hopefully as long as the home! Part of the durability decision will be determined by the climate in which you are building. Local materials are often an appealing choice since they provide a connection to the region and reduce transportation costs. You might find that you can use local stone for foundation or pier veneers and retaining walls, or local cedar siding or pine board and batten for the walls, if these materials are available.

Utility Spaces

Utility spaces support the more obviously "homey" spaces like the kitchen, living room and bedrooms. Put careful thought into their design, functionality and placement, and they will reward you by helping to make everything else function efficiently. Consider accessibility and proximities to other rooms and most common uses. It is a great idea to plan utility spaces early in home design planning; if they are placed as afterthoughts, they are less likely to truly enhance and support other activities in the home.

Here are some types of utility spaces to think about and plan for, depending on your needs:

- Mechanical room
- Basement
- Garage
- Mudroom
- Laundry room
- Closets
- Pantry

The mechanical room, home of the furnace or heat source, typically resides in the basement—in the open if it's an unfinished basement, or tucked away in a closet or small room if it's a finished basement. Mechanical rooms can require fresh combustion air or direct outdoor access for exhaust, depending on the heat source and type of fuel. If the mechanical room is closed off, it can make sense to have your utility lines come into this room in order to reduce clutter, and the room can be insulated to reduce noise as well. If you want to run utilities into a basement mechanical room, and the utility lines follow the driveway into the site, then the corner of the house closest to the utilities is the most efficient location. I strongly recommend that you do not place a mechanical room below or adjacent to sleeping or private areas of the home—it is best to build it under or adjacent to other utility spaces. Another way to optimize its placement is to make sure it is close to the center of the distribution of power and water, which is especially important for heating. It is more energy-efficient to distribute heat in close to equal distances throughout the home.

VENTILATION REQUIREMENTS FOR MECHANICAL SPACES
PROVIDING FRESH AIR TO MECHANICAL SPACES IS ONLY REQUIRED WHEN FUEL IS BEING BURNED, SUCH AS NATURAL GAS OR FUEL OIL. SYSTEMS SUCH AS GEOTHERMAL HEAT PUMPS OR COMPLETELY SOLAR-POWERED SYSTEMS DO NOT RELY ON FUEL COMBUSTION AND DO NOT REQUIRE FRESH INTAKE OR EXHAUSTS. DIFFERENT SYSTEMS FOR HEATING AND VENTILATION WILL BE DISCUSSED IN THE SECTION ON BUILDING SYSTEMS.

BASEMENTS ON STEEPER SLOPES CAN PROVIDE MORE USEABLE LIVING SPACE AND EVEN HAVE AN ENTRANCE FROM THE OUTSIDE AT THE LOWER ELEVATION WHERE THE WALL IS EXPOSED. IF YOU HAVE AN EXPOSED WALL, YOU CAN USE A CONVENTIONALLY FRAMED WALL INSTEAD OF CONCRETE.

Basements are commonplace in northern climates where foundation walls have to go an average of 4 feet (1.2 meters) into the earth to resist the way frost heaves soil. While it's more expensive to excavate deeper than that and add more concrete for full walls, the cost is small for the overall square footage gained. Today's basements are not as they once were. Advances in insulation and waterproofing, in combination with good construction practices, mean that you can have a basement that feels no different than the floors above ground, providing the same comfort, warmth and livability. These spaces can be additional bedrooms, workshops, media rooms or family rooms.

Garages can be anything from simple, unfinished structures to fairly elaborate affairs connected at the first and second story of the home with habitable space on the second floor, like a family room or in-law suite. Besides being used to keep vehicles out of the elements, garages also usually store household and yard items, two functions that should be considered in their design. Regardless of what exact form they take, garages pose unique design problems because their large doors reflect the scale required for their function. They do not match the aesthetics of the rest of the home which is designed for a more human scale. Barn or carriage-style garage doors can help reduce the perceived larger scale. While budget restrictions usually dictate a short driveway, it may be aesthetically preferable to have a more circuitous and indirect access to the garage to put the doors in a less visually prominent position.

The connection between the garage and home requires attention. If the garage is separate, enclosed access to the house from the garage may not be necessary; a covered or trellised

The enclosed connection between the garage and home is an ideal location for this home's sunroom and an adjoining breezeway.

Deeply recessed doors and a pent-roof give scale and aesthetic appeal to the garage-side of the house.

walk may suffice and be more aesthetically pleasing than a solid connection. Breezeways, mudrooms or sunrooms can offer a physical connection between the home and garage. Mudrooms provide an informal entrance to the home and space for storing outdoor garments. Mudrooms can also incorporate laundry facilities because they are where the most heavily soiled garments are often shed. If a mudroom or sunroom is designed to contain plumbing for laundry or sinks, it should be fully enclosed and heated at least minimally to protect pipes from freezing, especially in colder climates. A mudroom will need a floor that is durable and easily cleaned. Consider storage requirements, such as cubbies and closets that will affect its dimensions.

Laundry rooms can range in size and function from closets large enough to fit apartment-sized stacked washer/dryer units to large rooms with cabinets for storage, counters for folding and sinks for pre-washing. Another consideration for laundry rooms is their proximity to bedrooms and bathrooms and what floor they are on. Your laundry room should be close to where the

CARBON MONOXIDE SAFETY

CARBON MONOXIDE IS PRODUCED WHENEVER FUELS ARE CONSUMED. PLACE CO MONITORS ON EACH FLOOR LEVEL AS WELL AS IN GARAGES AND MECHANICAL SPACES. MAKE SURE THE GARAGE IS FULLY SEALED FROM ATTACHED HABITABLE SPACES, AND MAKE SURE THAT CONNECTING DOORS ARE WEATHER-STRIPPED PROPERLY.

laundry piles up, for example, near children's bedrooms or close to the regularly used entrance of the home. If you will have a tendency to do laundry in the evening, consider if sound-proofing is required in relation to adjacent bedrooms.

There never seems to be enough storage space. Think of where closets and pantries can be conveniently located and what things they should be designed to hold. For example, a pantry will typically be used to store dry goods and cans, but it may also need room for a wine rack, a space for less often-used kitchen appliances or a bin of pet food. For each storage space, decide whether it will be small front-access only, or a walk-in. Decide where the shelves or bins will be and the heights and sizes of storage spaces that will therefore be required. Plan for logical and convenient access—the linen closet should be near the bedrooms and bathroom, and the pantry should be adjacent to the kitchen. If space is at a premium, think about maximizing typically wasted spaces that exist under stairs and eaves, using built-in shelves and cabinets.

Public Spaces

Public spaces are varied in size, purpose and style, but have in common that they are designed for use by everyone who spends time in your house, residents and guests alike. Public spaces to consider when designing your home can include:

- Entry or foyer
- Stairs
- Kitchen
- Dining room
- Great room
- Living room
- Media room
- Lofts, balconies and galleries
- Sunrooms

We have previously discussed the entry or entries into a home in relation to street or vehicular access, as well as in relation to garages and mudrooms, which are typically associated with informal entrances. Decide if you want both formal and informal entrances, and how they will be used. Keep both residents and guests in mind as you look for the answers to these questions: Do they provide enough space for the number of people you expect to use them? How about storage spaces for outdoor clothing and footwear? What rooms do your entrance spaces lead into? Do your entrances offer views into adjoining spaces, beckoning visitors forward and into your home? How do you want people to experience the first space in your home that they enter—warm, grand, inviting, or intimate?

Stairs can connect from public spaces to private spaces or to other public spaces. From a space-use perspective, stairs are most efficient when placed one flight over the other. It's best to put stairs in a central location or near the highest-traffic areas on each floor. For example, if the kitchen, living room and foyer are adjacent to each other on the first floor, the juncture between them might be a fitting spot for the stairs because it will require less travel and hall space.

Kitchens were traditionally more of a private work space and not the hub of social activity that they have become in today's homes. People tend to gather in kitchens more often than other rooms, making them a common focal point for family activities and larger gatherings. Placing the living room and dining room adjacent to the kitchen is common, and if you choose an open floor plan, these rooms might even flow together, allowing for gatherings to remain cohesive throughout. For the kitchen, consider how much you like to cook, what the typical workflow will be like, what kinds of appliances and storage space will be required, and the views into and out of the room.

Decide if you want a breakfast nook or smaller eating space close to the kitchen to provide an area for informal meals. Because so much time is spent in a kitchen and adjoining dining nook, light and orientation may be especially important to you. If you like to read the morning paper while having a cup a coffee or tea, then these spaces could be oriented to receive morning sun.

For entertaining, a deck adjacent to the dining room allows for easy flow from one space to the next.

Alternatively, if cooking a late dinner for two while sharing a bottle of wine is more your preference, you might want to put the kitchen and informal dining nook on the western side of your home.

Dining rooms can be either separate rooms or part of a larger open space. If you have a more open floor plan, you can define the dining area with lighting, a different floor pattern, or even a large area rug. Dining spaces can have cathedral ceilings or lower, more intimate ones. Whatever you choose, lighting is always important to get right—it must be primarily focused on the table and illuminate the space well, but not be too bright.

If living rooms or great rooms are part of an open plan, they should be adjacent to kitchen and dining areas. Great rooms are typically defined as being cathedral spaces, while living rooms have a flat ceiling above them which is part of the second story of the house. Great rooms and living rooms can serve many purposes but the two most common are for entertaining and watching television. It may be a challenge to balance the differing spatial needs that these two activities demand with other variables like views to the outdoors and the location of audio/video equipment and fireplaces or woodstove. Consider the size of the space, the frequency of its uses and then consider transferring some of them to other rooms. For example, you might want to move the television and entertainment center to a family or media room, leaving the living room primarily for entertaining and reading.

Home theaters and media rooms have become increasingly popular. Media rooms are designed specifically for viewing movies, sports and television shows. You will probably want to ensure

Natural light, materials of light color and an open design make this kitchen a focal point of this home.

This room has a number of different areas, each of which offers comfortable, intimate spaces for small groups of people.

optimum sound quality, which means attention must be paid to room layout, reflective and absorptive surfaces and the isolation of the media room acoustically from other rooms in the house.

Lofts, galleries or balconies are typically open on one or more sides to the space below, with half walls or railings and balusters for safety. These types of spaces can make wonderful reading areas, office spaces or even small libraries. A balcony might provide a hall in front of several bedroom doors, creating a cozy transitional space between private and public spaces.

Sunrooms can be seasonal with no insulation, inexpensive single-pane glazing and no heating or cooling; or fully insulated and conditioned as part of the home. A sunroom can also be combined with a screened porch area. An insulated and heated sunroom with windows that open and are placed on the south or south-east side of the home can function in colder months as a warm and well-lit space for brunches or reading; and in summer months serve as a cooler place to relax with a drink in the evening.

As well as being functional, balconies can provide visual interest without detracting from an open concept design.

A sunroom can be seasonal, or it can be designed for use in the colder months as well.

Private Spaces

Private spaces to incorporate into the plan for your home can include many or all of the following:

- Master suite
- Master bedroom and bathroom
- In-law suites
- Bedrooms
- Bathrooms
- Den/office/library

Master suites generally consist of a master bedroom, private bathroom and a walk-in closet. Make sure that the bedroom is at least large enough for the bed, nightstands and dressers and possibly a sitting or reading area. The master bathroom typically contains a combined tub and shower unit or separate shower and tub (sometimes a Jacuzzi-style), a toilet, a sink or sinks and possibly a bidet. Many couples prefer to each have their own sinks. Depending on the choices of fixtures, the master bathroom can be quite large. The walk-in closet can be situated off the master bathroom or

bedroom and typically has between 10 and 30 linear feet (3 to 9 meters) of hanging storage (which may, of course, be stacked).

In-law suites are similar to a master bedroom suite. They may contain small living and kitchen spaces as well. An in-law suite might have its own entrance from the outdoors and a small entry space for storage.

Individual bedrooms can be designed for specific occupants, such as children or guests, but it's often best if the design is versatile enough that you can reassign bedrooms over the years. Obviously, a bedroom must be big enough to accommodate a bed, nightstand and dressers and allow circulation space around these things. The room should also have a closet of adequate size for its expected use. It's great to give bedrooms windows on at least two walls, which will allow for ample light and excellent cross-ventilation.

Bathrooms will vary in size dependent on ancillary uses. A first floor bathroom might be just large enough for toilet and sink, or might also contain a shower/tub unit, or even a washer/dryer tucked in a closet. Second floor bathrooms will typically have toilet, sink and shower/tub. The design and location of each bathroom will depend on who will use it most often and what it is close to. Bathrooms always require ventilation and should always have natural light from at least one wall.

Dens, offices and libraries require isolation from the louder, more public areas of the home. If you want a home office, you might consider adjoining it to the master bedroom, creating a truly private environment. Offices might also be located adjacent to other quieter areas of the home, such as a guestroom or a second floor loft. Home offices should have a good measure of sound isolation which can be accomplished through either physical separation or sound-attenuating insulation. A home library meant for book storage and reading might more readily be placed in a lofted space that is more open to its adjacent rooms.

Miscellaneous Considerations

Now that you've thought about all of the different rooms, there are a few miscellaneous but important aspects of design to consider in some detail. These include:

- Interior materials and finishes
- Windows
- Doors
- Fireplaces, woodstoves and masonry heaters

A central fireplace or hearth has an enduring appeal and creates a kind of anchoring point within the home.

Interior materials and finishes account for a large portion of the overall budget and will affect both the architectural style of the home and the quality of the indoor environment. Avoid using materials that need to be replaced or refinished frequently. Paints, carpets and plastics release or off-gas volatile organic chemicals (VOCs) into the indoor air, which at times cause unpleasant odors and have been linked in some studies to possibly harmful physical effects. Local stones and woods are good choices for flooring materials. Clay plasters and low or no-VOC paints are also good choices for wall and ceiling finishes if you are concerned about chemical off-gassing throughout the life of the home and indoor air quality.

Windows admit light and allow for ventilation throughout the home. Windows are also a sizeable investment and should be carefully weighed into your budget. The most common types of windows are casement, double-hung, awning or sliding. These can be made from many different types of materials, including wood, fiberglass, vinyl or aluminum. As for the style of window, this will largely be based on the aesthetic of the house. Sliding windows are less efficient than casement windows from the perspective of air filtration, but most windows today are built well and within acceptable ranges of efficiency. Do your research to make sure you invest in a quality product.

Sliding and double-hung windows have two panes and are opened by sliding one in front of the other. For sliding windows, the sliding pane moves horizontally, while double-hung windows move up and down.

When closed, casement and awning windows typically seal more tightly and allow less air seepage than sliding and double-hung windows. When open, casement and awning windows allow more ventilation than sliding or double-hung windows of the same size.

Doors with large amounts of glass are inviting and allow light to pass through, illuminating the spaces inside and providing views in both directions. Solid doors or doors with smaller or glazed windows suggest security and privacy, and are appropriate for entrances that are visible from the street or other public areas. For any kind of door, consider how it will appear from both the inside and outside of the home, and how the entrance to your home will be perceived.

Most will agree that open fireplaces are very romantic, but unfortunately open fireplaces are very inefficient at generating heat and in most cases they will actually remove heat from a home. Quality construction, dampers, glass doors and air supply from the outdoors for combustion will help increase a fireplace's efficiency if you decide to have one. Woodstoves and masonry heaters are both excellent options if you wish to burn wood for heat. Woodstoves operate much more efficiently than a fireplace and are available in materials like cast iron, steel and soapstone combinations, and can even be purchased in different colors. Wood stoves may have a full masonry chimney made of brick or stone, or simply an insulated steel pipe. Masonry heaters are similar to woodstoves in that they are a closed burning environment; however the major difference is the flue of the stove loops several times through the large masonry mass before it exits the stove. Masonry heaters are usually made of soapstone, and the looping flue allows the hot combustion gasses to heat the stone most efficiently. Once the fire has heated the stove, the stove will radiate the heat from the small fire for hours after it has been extinguished—sometimes up to twelve!

Now that you have defined your vision, it's time to move on to the formal design process, where all this groundwork will really pay off.

TYPES OF WINDOWS

CASEMENT AND AWNING WINDOWS OPERATE WITH A CRANK-STYLE HANDLE THAT OPENS THE WINDOW OUTWARD. CASEMENT WINDOWS PIVOT FROM THE RIGHT OR LEFT; AWNING WINDOWS PIVOT FROM THE TOP SO THAT THE WINDOW SWINGS UPWARDS AS IT IS OPENED.

chapter three   03

PRE-DESIGN

"NOTHING IS AS DANGEROUS IN ARCHITECTURE AS DEALING WITH SEPARATED PROBLEMS. IF WE SPLIT LIFE INTO SEPARATED PROBLEMS WE SPLIT THE POSSIBILITIES TO MAKE GOOD BUILDING ART." **– ALVAR AALTO**

KEEPING PERSPECTIVE ON THE DESIGN PROCESS

In order for the perfect design to emerge, it is ideal to view the entire design development process holistically, and avoid "dealing with separated problems", as Alvar Aalto suggests. It can be difficult to keep the big picture in mind because of the need to develop your design in increments and stages with all the necessary attention to detail that was discussed in Chapters 1 and 2. However, remember that each part is interdependent with other parts of the design; each decision you make should be carefully considered in regards to its impact on the design as a whole. As an architect, I find this to be one of the most challenging and rewarding aspects of the design process—balancing the characteristics of the different parts and bringing them together to create a harmonious whole.

This 'iterative' process that I mentioned earlier requires you to repeatedly pull apart your design plans, analyze them, and then reassemble them. There are usually multiple outside parties providing input, including the architect and builder, engineers and interior designers. While so many participants can make design discussions more complex, these different parties will contribute valuable information and expertise. It is time-consuming and sometimes challenging to manage the flow of information and to consider how to apply it to the design, but it is also important. The reward will be a house that you love to call home.

PHASES OF DESIGN DEVELOPMENT

The commonly recognized phases of architectural design are pre-design, schematic design, design development and the preparation of the construction documents. In most instances they serve as milestones in a design schedule leading to a construction deadline. If each milestone is met, the design is considered complete and ready for bidding, to be followed by construction. These phases help structure the design process and help everyone involved monitor its progress. While the process is defined in a linear fashion, it is common to revisit aspects of earlier phases periodically in order to re-evaluate decisions and to make sure that the design is still aligned with the vision you defined in Chapter 2. This chapter focuses on the many aspects of pre-design; regulatory guidelines, building systems, heating and cooling systems, building shells, lighting and power, project scope and responsibilities, project schedule, project budget and contingencies.

ABOUT PRE-DESIGN

A big part of pre-design is "programming", which is the process of taking all the important groundwork you did in Chapter 2 and formally documenting it with your architect. The formal outline should include information about your aesthetic preferences in architectural style; desired site characteristics; required spatial relationships and important functional adjacencies; as well as details about site and environmental conditions. At this stage you and the architect can further refine your outline by organizing rooms and spaces by story, which will help determine the approximate square footage required on each floor.

Pre-design also involves the identification of applicable codes, regulations and zoning issues as well as the consideration and preliminary selection of building systems (such as heating/cooling and electricity). Additionally, the scope of the project, consultant and contractor responsibilities, project phasing, project budget, budget contingencies and schedule should be established in this phase. This chapter will focus on how these other considerations can affect the development of your design. With the outline and vision formally in place, you should be in a good position to make all decisions—whether they are about design, schedule, budget or anything else—in accordance with your priorities.

UNDERSTANDING YOUR SITE

Your building site provides the context for your design. The first step in developing a formal design is to document and review the site characteristics and existing site conditions, while the second step addresses the vision of the site as it pertains to the design of the home. This may or may not require the involvement of professional consultants. A surveyor can provide drawings of the site if none are available from your local city or town hall. Additional drawings may not be necessary depending on the complexity of the site or the presence of previous survey work. However, basic site drawings must be reviewed in order to assess such physical characteristics as slope, vegetation, soils, ledge, existing bodies of water like ponds, brooks, or vernal pools, as well as existing site utilities or structures.

Next, consider the character of the site itself in its environment: the neighborhood. What are the characteristics

If a site has enough slope it will allow for a walk-out basement, which can give you even more functional space and allow significantly more natural light to enter.

of other homes or buildings in the area and their sites? This is sometimes referred to as the neighboring vernacular. Are there historic precedents that should or must be addressed?

The second step is to prepare a more technical outline of the site. The survey of the site should document as many of the following as possible and applicable:

- Topographic analysis of the site within the context of its neighbors
- Slope analysis (grades)
- Physical features
- Trees, vegetation
- Rocks, ledges and soils (referring to geotechnical reports if they are available)
- Bodies of water
- Vehicular access
- Utility access
- Existing structures, roads and utilities
- Prevailing winds
- Solar access

The outline of the site design can be intrinsically tied to your vision for your home or it can be done separately. In either case, the key points to address in the outline are as follows:

- What is the neighboring vernacular?
- What issues of visual or aural privacy are present?
- Does the site offer views distant or local to the site?
- How does the sun affect the site?
- What are the primary relationships or adjacencies between site and home?

For each of these questions, ask yourself the secondary query of "…and how should the design respond?"

Once you have the existing site conditions documented and an outline prepared to guide the design of the site developed in tandem with the design of the home, you can then move on to research and address technical options about the site.

SITE ACCESS

It makes sense to look at site access first, since it will be needed throughout the process from the beginning of site work to the end when the home is built and occupied. You may choose to leave an existing driveway or decide to shift its location to suit your functional or aesthetic requirements.

Look at site access in tandem with utility access. Often, the most feasible option is to run the two in proximity to each other, especially if utilities will be brought to the house underground. Note that access must meet any applicable local zoning requirements, especially in regards to distance from adjacent properties, corners, hills and drainage. Finally, consider the other impacts that site access will have; for example, the larger or longer the driveway, the greater the disturbance that will be caused to the site, and the more materials that will be required. Whether you choose asphalt, concrete or a semi-permeable surface, the water that will run off of the drive must be addressed with the use of swales or undersurface drainage like drywells.

UTILITIES

Let's talk now about utilities, since site access and utilities access can be planned in tandem. These include power, cable, telephone, potable water and waste water removal.

Power, Cable and Telephone

Power, cable and telephone are usually run along streets on utility poles. Service can then be "dropped" to the home directly from a pole. This option limits the distance a home can be from a pole because there are minimum heights that utility wires must be from the ground. If the desired location of the home is too

Careful planning allowed for many of this site's trees to remain standing.

distant to connect directly from the pole, you can either install additional poles on the site or you can bring the utilities to the house underground. Both of these options are at your expense. While underground delivery of utilities is usually more expensive, the benefit is that there will be no poles on the site to impede views and you eliminate the possibility that trees on the site will damage the lines.

City Services: Potable Water and Waste Water Removal

Utilities typically include city-provided services such as potable water and waste water removal, both of which are underground. Potable water is metered most commonly in the home at the point of entry, typically a basement or crawlspace. If waste water removal is provided it can be either gravity-fed from the house or an ejector pump might be required depending on the grade of the lot. It can be gravity-fed if the outlet from the home is above the inlet into the city line, based on the minimum slope required and the distance between inlet and outlet. An ejector pump simply pumps the waste water up to the city line.

Wells

If no potable water is provided, you will need to dig or drill a well. Dug wells are commonly shallow wells lined with stone or concrete. Water flows into the well and is then pumped into the home. Drilled wells are done with specialized equipment. They can be shallow or up to several hundred feet (or meters) deep. Drilled wells are lined with concrete, PVC or steel pipe. Water enters at the bottom of the well, flows into the pipe and then is pumped into the home.

Septic Systems

If municipal waste water removal is not provided, you'll require an on-site septic system. An engineered septic system is often required by building codes, which means that you will have to hire an engineer or certified designer, depending on your local regulations. There are several types of systems available, all of which need to be sized according to the number of occupants in the household and the types and depth of soils on the site.

The location of any septic system is important. It is common to specify a few possible locations on the site plan that are determined by the preferred location of the home and the system in relation to one another. Typically, test pits have to be dug to verify soil depths and percolation rates to determine what is feasible and what kind of system will be required. Septic systems do not last forever and so accessibility for maintenance and replacement should also be considered when planning site access and the septic system location in relation to the home.

Gravity and pump-to-gravity systems are the two most common basic septic systems. They have in common that effluent is drained from a septic tank to a drain field. The drain field is a series of perforated pipes laid over gravel-lined trenches that are usually 2 to 3 feet wide (60 centimeters to 1 meter) and have 3 to 4 feet (1 to 1.2 meters) of mostly dry, undisturbed, permeable and well-oxygenated soil beneath the trenches. The effluent is treated by natural chemical and biological processes as it percolates down through the gravel and soils so that the effluent is clean by the time it reaches groundwater. The only difference is that the gravity system relies on the drainage field being below the level of the septic tank so that gravity causes the effluent to flow from the tank to the field. The pump-to-gravity field system employed where the gravity system won't work on a site, means that a pump is used to move the effluent from the tank to the drain field.

When optimal soil depth or quality is not available or when greater efficiency is required, you can add a pressure distribution system to one of the aforementioned septic systems. A pressure distribution system uses a pump and pressurized lines to ensure the even distribution of effluent throughout the drain field. In this way the effluent is allowed to percolate more evenly throughout the drain field and while the tank fills again.

> **CONSERVING WATER AND SAVING MONEY**
> THE TYPICALLY LOW ADDITIONAL COST OF INSTALLING WATER-EFFICIENT PLUMBING FIXTURES SUCH AS HIGH-EFFICIENCY TOILETS AND LOW-FLOW FAUCETS AND SHOWER HEADS WILL CONSERVE WATER AND SAVE MONEY IN THE LONG RUN. EVERY GALLON OF WATER THAT IS SAVED IN USE IS ALSO SAVED IN TREATMENT OR DISPOSAL. IN ADDITION TO BEING GOOD FOR THE ENVIRONMENT AND HELPING SAVE MONEY, WATER-EFFICIENT CHOICES CAN PROLONG THE LIFE OF YOUR SEPTIC SYSTEM.

A sand filter can also be used if you're dealing with insufficient soil depth. A sand-filled containment vessel is placed between the septic tank and a pressurized drain field. The effluent trickles through the sand where it is subjected to chemical and biological processes to clean it prior to being distributed to the field.

Yet another solution for inadequate soil depth is a mound system. A mound is a drain field that is raised above ground level and made of sand fill material specifically for use in septic systems. Mound systems are typically used in conjunction with a pressure distribution system. Proper drainage around mound systems is critical, especially on sloping sites.

REGULATORY GUIDELINES

Building codes, zoning regulations and specialty regulations exist in the interests of the health, safety and welfare of the public. Building codes will vary from province to province, state to state, county to county, and sometimes even from town to town. For this reason I recommend that you find out what the specific regulations are in your area. However, the following section presents some of the issues that you will need to consider in regard to these regulatory guidelines.

ZONING REGULATIONS

Zoning is a form of land-use regulation intended to prevent new development from negatively affecting existing established land-uses as well as segregating uses that may be incompatible. Zoning is typically enforced by county or municipal governing bodies.

Zoning regulations deal primarily with the permitted use of land, based on a zoning map of a community. Each zoned land-use, such as residential, commercial or industrial will also have density regulations, designating some areas for low-density housing like single family residences, and some areas for high density, such as high-rise apartments or condominiums. The most important aspects of zoning as they pertain to residential use are setbacks, minimum lot sizes, building size relative to lot size, building heights and the proportions of the different types of space on the lot.

Setbacks determine the required minimum distances from the street frontage, from side lots (neighboring properties) and from the rear lot line (which might also be neighboring properties). In more rural areas, setbacks may be substantially greater than those for an urban or suburban setting. Setbacks may also include utility easements or rights-of-way through the site, which are the rights reserved by utility companies to run their pipes and wires through designated paths on private property. Minimum lot size also limits the sub-division of larger lots, which helps control the density of lots and homes. The building size compared to lot size is a third way to control density. It might be that a home can only cover one quarter of the lot, in this way controlling the footprint of the home and preserving a certain amount of open space. Building height is also sometimes regulated and expressed in measurements above grade (above ground). For example, if the maximum height is 35 feet (10.5 meters), a simple gabled home would be allowed to have its highest ridge no more than 35 feet (10.5 meters) above grade at any point. Finally, regulations about ratios of different types of spaces on a single lot should be examined. Some zoning

Special attention to town, state or municipal regulations must be paid to sites that border bodies of water.

Maximizing day-lighting is the easiest way conserve energy and improve the quality of the indoor environment.

regulations will have maximum and/or minimum percentages of spaces such as the area of the driveway in relation to structures like the house and garage, or paved areas in relation to open vegetated areas like lawns, gardens, and woods.

BUILDING CODES

Building codes address the minimum acceptable level of safety for structures; issues pertaining to health, safety and welfare. In the United States, where the regulation of construction and fire safety lies with the local authorities, the system of model codes is used. A model code is typically written by a non-governmental entity and is then adopted by the authority that has jurisdiction. Once adopted, this code becomes law by the practice known as "adoption by reference". The code is then used by architects, engineers, safety inspectors and other bodies such as developers, contractors and manufacturers of building products.

Building codes address issues of structural and fire safety; the ability of a building to resist natural forces such as wind, snow and earthquakes, and its capacity to protect occupants and neighbors in the event of fire. Energy conservation is currently a point of concern. The construction industry taxes natural resources and fossil fuels heavily, both in the construction phase as well as in the operation of a building; for example, its heating and cooling, its supply and disposal of water and wastes. Building codes identify health requirements like indoor air quality standards, plumbing standards, minimum room sizes and natural light. The quality and installation of building materials is also addressed, as well as the qualifications of the people doing the work.

Building codes are currently more prescriptive in nature, indicating exactly how and of which materials a building is to be constructed. It is becoming more common for performance-based codes to be used. A performance-based code indicates the desired result of how something must perform (for example, how fire-retardant building materials must be) and it is left to the designer to prove how this will be accomplished. Performance-based or prescriptive, building codes are designed for the public's protection and most often will not address issues of aesthetics.

SPECIALTY REGULATIONS

Specialty regulations can vary widely from jurisdiction to jurisdiction. Counties or municipalities might have regulations concerning indigenous or native vegetation, local environmental concerns or light pollution requirements. There might even be specific contractual regulations concerning parking, material and color selections, and even architectural style or landscaping requirements.

Some regulations pertain to specific geographical and regional characteristics such as earthquakes, hurricanes or wildfires and specify the minimum performance that the building must withstand in those conditions. For example, in areas subject to wildfires there are minimum distances specified between home and vegetation like shrubs and trees, as well as strict requirements pertaining to the flammability of exterior materials.

Housing developments may impose regulations on the design of buildings. These guidelines are set and agreed to when a lot is purchased with the intent to build. These types of guidelines are most often established to maintain a particular type of aesthetic throughout a development. For example, the design board of a mountainside resort development might limit the overall length of roof ridge, or request that all exterior siding selections and colors be submitted for approval. By requiring that all colors be approved, the design board can control the color palette of the development, perhaps choosing muted earth tones to blend with the surrounding forest and mountain; by restricting ridge length, they can limit the scale of unbroken vistas of roofs.

BUILDING SYSTEMS

Now that you have mapped out your site and its characteristics, planned your site and utilities access, and investigated local building regulatory guidelines, you can start planning your home. Let's start at the heart of your home, with the building systems. Building systems are the engines that run a home, including electricity for lights, outlets and appliances; water for sinks and showers; and furnaces and boilers for supplying heat—to name just a few. These systems will vary in their complexity as well as their impact on your budget. Some common system designs are done by the supplier of the materials or the contractor installing them. Other systems require the expertise of a mechanical engineer at this stage of the project.

Remember that it is best to consider these systems together as you develop your overall design. What you choose for fuel will affect the type of heating and cooling system you select, as well as how you will construct the building shell and implement ventilation. The main benefit in planning these things together is that you will be able to optimize each in relation to the other—significantly increasing the efficiency of their implementation and use, and reducing costs.

FUEL OR ENERGY SOURCE

The building's primary fuel source is one of the first and most important decisions to make. Common fuels are heating oil, natural gas or propane gas. Deciding which fuel source to use will usually be based on availability, cost or personal preference. You will have already identified the availability of local fuel sources from your municipality in the site survey stage. Cost should be assessed by considering both the costs involved with setting up the system, as well as operational costs over time. It is common for more efficient systems to have slightly higher initial costs, but these are balanced by lower operating costs over time. Discuss these issues with your architect, builder and engineer or heating/cooling system designer, so that specific regional costs of installation, fuels, comfort and environmental impact can all be considered.

NATURAL GAS

Natural gas is a gaseous fossil fuel primarily made up of methane. In more urban areas it is typically available at the street and piped to the house where it can be used for boilers and furnaces to provide heat as well as to supply gas fire to ovens and stove tops. In more rural areas, natural gas can be supplied as compressed natural gas (CNG) in large tanks above or below ground.

PROPANE

Propane is similar in its use as a domestic fuel source for heating the home, heating hot water and supplying fuel to the oven and stove. Propane is usually derived from petroleum products during the processing of oil or natural gas. Like natural gas, it can be compressed into a liquid to make it easier to transport. When compressed it is commonly referred to as liquefied petroleum gas (LPG or LP gas).

OIL

Oil or fuel oil is another option. Widely available, fuel oil is a liquid petroleum distillate. Fuel oil is most commonly stored in a tank in the basement and is delivered to the home as a liquid in trucks.

REDUCING ENERGY CONSUMPTION

ENERGY USE CAN BE REDUCED IN MANY WAYS. TAKING ADVANTAGE OF NATURAL LIGHT WITH THE STRATEGIC LOCATION OF ROOMS AND WINDOWS REDUCES THE NEED FOR ARTIFICIAL LIGHTING. COMPACT FLUORESCENT BULBS LAST MANY TIMES LONGER AND USE MUCH LESS ENERGY THAN INCANDESCENT BULBS. YOU CAN ALSO PURCHASE ENERGY STAR RATED APPLIANCES INCLUDING REFRIGERATORS, DISHWASHERS, WASHERS, DRYERS AND HOT WATER HEATERS THAT WILL REDUCE THE ELECTRICITY CONSUMPTION OF A HOME.

A balance of natural and artificial lighting is important for spaces to function properly at different times of day.

Electricity

Electricity is a very refined power source and its best application in a home is to power lights, motors and appliances. Although it can be used to heat a home through the generation of electrical resistance heat, it is not efficient and is very costly. When electricity is used for heat it should be in the most efficient way possible, such as to power a heat pump, which will be discussed in more detail in the next section. In short, the use of electric resistance heat for heating a home should be your last option if your home is in a climate that requires more than very occasional periods of heating.

HEATING AND COOLING SYSTEMS

Common heating systems include forced air and hydronic systems like baseboard radiators or radiant floor heat. All three of these systems can be run from boilers and furnaces fueled by any of the fuel choices mentioned in the previous section. Heat pumps are growing in popularity and are another system to consider. Cooling systems include air conditioning and fans.

Forced Air

Forced air systems require a furnace that burns fuel to produce hot air that is blown into the home through ducts and vents. Once the temperature drops below the setting on the thermostat, more fuel is burned to produce more hot air that will mix with the cooler air of the home resulting in a pleasant ambient temperature.

Forced hot air has two main benefits: the home's temperature can be changed dramatically in a short period of time; and an air conditioning system can be incorporated to make economic use of the heating system's blower and distribution ductwork. However, changing the heat of the home rapidly requires disproportionately greater fuel consumption, making for a less efficient system. The second drawback is air quality in the home. Because air is constantly being disturbed, dust and particulates in the ductwork must be filtered out and filters and ductwork cleaned regularly. Forced hot is one of the lowest initial first cost heating options.

Hydronic Systems

Hydronic systems include baseboard radiators or under-floor radiant systems that require a boiler to heat water. Both types of hydronic systems can use any of the three fuels discussed in the section above. The heated water is then transported to where it will radiate that heat throughout the house.

The most common method of transporting heat throughout the home is through baseboard radiators. Current baseboard radiators consist of a copper pipe with many closely spaced aluminum fins fitted over the tubing. Heat is transferred to these fins through conduction, and air passing through the fins carries the heat into the room through convection. Baseboard radiators are clean and quiet. The main drawback to this heating system is a slower response time than a forced air system in heating the house suddenly. Its second drawback is that if you decide to have air conditioning, then a separate system must be installed, including ductwork.

Radiant floor heating systems deliver the heat from the boiler through flexible piping either directly under the floor

---

GETTING THE RIGHT-SIZED HEATING OR COOLING SYSTEM

IN TIMBER FRAME HOMES CONSTRUCTED WITH SIPS, THE BETTER INSULATION AND AIR-TIGHTNESS OF THE HOME ALLOWS YOU TO USE SMALLER HEATING AND AIR CONDITIONING UNITS, ESPECIALLY IF YOU HAVE CHOSEN NEWER AND MORE EFFICIENT HEATING TECHNOLOGIES SUCH AS RADIANT HEAT OR GROUND SOURCE HEAT PUMPS. MAKE SURE THAT THE HEATING AND COOLING DESIGNER ACCOUNTS FOR SOLAR ORIENTATION AND HOW YOUR HOME HAS BEEN CONSTRUCTED; A SMALLER AND MORE EFFICIENT SYSTEM WILL OFTEN PERFORM BETTER AND LONGER THAN AN OVERSIZED SYSTEM AND IT WILL HAVE A LOWER FIRST COST.

Dormers can provide comfortable alcoves with lots of natural light.

system or embedded in the floor system. In the first instance the piping can be installed directly below the subfloor, commonly fastened to its underside in a reflective aluminum sheet designed to radiate the heat upwards. Below that, the rest of the floor cavity is usually filled with fiberglass batt insulation to aid in directing the heat upwards. The second common method is to encapsulate the tubing in a lightweight gypsum concrete. This method is more thermally efficient, providing better thermal mass to retain and deliver the heat to the spaces, but it has a higher first cost and typically will add a few inches of depth and more weight to the floor system. This will need to be accounted for in the detailed design of the structure. Radiant systems are extremely comfortable. The heat is stratified so that if the thermostat, located approximately 5 feet (1.5 meters) above the floor, is set to 70°F (21°C), the temperature at the floor might be as much as four degrees warmer, giving the space a

USING SOLAR ENERGY
FREE LIGHT MAY BE THE BEST AND MOST SIMPLE USE OF SOLAR ENERGY, BUT THE SECOND BEST IS TO USE IT TO HEAT WATER FOR YOUR HOME. TO HEAT WATER, HAVE SOLAR COLLECTORS INSTALLED ON THE ROOF OF YOUR HOUSE OR ON THE SITE SOMEWHERE NEAR THE HOME. THE COLLECTORS ARE FILLED WITH AN ANTI-FREEZE FLUID THAT RUNS INTO THE HOME AND THROUGH A HEAT EXCHANGER. THE HEAT EXCHANGER THEN HEATS WATER THAT CAN BE USED TO HEAT OR SUPPLEMENT THE HEAT SOURCE OF THE HOME.

This gabled dormer centers the space, provides the necessary headroom, permits ample natural light during the day and offers great views while bathing.

feeling of being warmer than it is. The drawback is that if you want air conditioning, a separate system of ducts will have to be installed.

Heat Pumps

Heat pumps can be ground source or air source. One type of air source heat pump that is used for cooling is a condensing unit that is placed outside of the home. It works by removing the heat from inside and expelling it outside. Ground source heat pumps work in a similar way, but they transfer the heat to and from the ground at a depth where the earth has a constant temperature. The benefits of ground source heat pumps are that they are highly efficient, eliminate the need for fuel storage or supply on site, offer extremely low energy costs, and can work in both heating and cooling modes. Heat-pumps are growing in popularity. I believe that they will increasingly take the place of boilers and furnaces for residential heating and cooling applications, due to increasing costs of fossil fuels. Heat pumps tend to work best in places that do not have severe winters.

Air Conditioning and Fans

Air conditioning can be combined easily with a forced air heating system, using the fan from that system. The fan will blow the cooled air throughout the home in ductwork and out through the registers. Because air conditioning systems use an external condensing unit that can make a low humming sound when in operation, the placement of this unit should be carefully considered from an aesthetic and noise perspective.

You can also consider using ceiling fans inside the home to help circulate air and create a cooling effect.

> **USING SOLAR SHADING**
>
> TO HELP KEEP YOUR HOME COOLER NATURALLY, YOU CAN USE SOLAR SHADING TO KEEP THE HEAT OF THE SUN OUT. ONE METHOD IS TO MAKE USE OF DECIDUOUS TREES NEAR THE HOME. THE CANOPY OF LEAVES WILL PROTECT THE HOME FROM THE HEAT OF THE SUN IN THE SUMMER, BUT ALLOW THE HEAT AND LIGHT OF THE SUN TO REACH THE HOME IN THE WINTER WHEN THE TREES ARE BARE. ROOF OVERHANGS CAN PROVIDE SOME SHADING, AS CAN AWNINGS OR OTHER SHADING DEVICES THAT ARE ATTACHED TO THE OUTSIDE OF THE BUILDING. YOU CAN ALSO USE SHADES INSIDE THE HOUSE TO HELP CONTROL UNWANTED HEAT GAIN OR LOSS.

BUILDING SHELLS

The efficiency of heating and cooling systems is tied to how well a home is insulated from the outdoor environment. This brings us to the next major consideration: the shell of the building—in other words, the walls and roof—as well as the holes we put in that system, primarily the windows and doors.

Materials used for the building shell are assigned R-values to indicate their resistance to heat flow. For comparison, a well-insulated wall will have a higher R-value in the mid-20s; a well-insulated window will have a much lower R-value in the single digits.

Basement Walls

Let's start with the basement walls or frost walls (a shallow space under the home) that can be below grade (underground). Building codes usually require that the concrete walls of a basement or crawl space be insulated from the cold earth surrounding and supporting it—and even if insulation isn't required, it's a good idea to put some in anyway.

One way is to put rigid foam insulation along the outside face of the wall, separating the wall from the earth. The same method can be used to insulate a basement concrete floor from the earth, which is very important if any habitable rooms are to be placed in the basement portion of the home. The relative temperature of the earth at that depth below grade is usually about 50 to 55°F (10 to 13°C), an uncomfortable temperature for most feet.

Another insulating method is to use insulated concrete forms (ICFs) to create the foundation walls for the home. ICFs are forms made of insulating foam, commonly sold in blocks that are fitted or fastened together. They are used as the form for the wall and then left in place as insulation on either side of the concrete after it has been poured. ICFs will typically have an R-value between 18 and 28, allowing for a substantially more comfortable interior environment. Most ICF companies also provide fastening strips within the form, which allows the direct application of furring strips and then an interior finish such as drywall. ICFs have a higher initial cost but reduce overall energy costs over the long term, as well as positively affecting the habitability of the space.

Walls and Roofs

You can either use conventional framing to enclose your timber frame house with walls and a roof, or you can use insulated panels. I believe in the superior performance benefits of insulated panels and would strongly recommend this option for timber frame structures. The benefits of insulated panels were briefly discussed in Chapter 1, in the sections about *Timber Frames and Insulated Panels* and *Construction Types*.

SIPs provide for extremely well-insulated buildings and allow for substantially less air infiltration and exfiltration, which contributes to their higher performance compared to conventional construction. SIPs are able to resist lateral loads, such as those caused by wind, as well as seismic and gravity loads like snow. SIPs resist these loads and transfer them to the timber frame, to the foundations or to another resistive structure.

DURABILITY OF BUILDING MATERIALS

SELECTING HIGH QUALITY AND HIGH DURABILITY MATERIALS CAN GREATLY REDUCE THE FUTURE EXPENSES OF A HOME. THIS APPLIES NOT ONLY TO EXTERIOR MATERIALS SUCH AS SIDING AND ROOFING, BUT TO INTERIOR FINISHES AS WELL AS APPLIANCES AND BUILDING SYSTEMS. TRY TO MAKE YOUR SELECTIONS BASED ON EXPECTED LONGEVITY, EASE OF MAINTENANCE AND IF IT CAN BE RECYCLED WHEN REPLACED. FOR EXAMPLE, ASPHALT ROOF SHINGLES ARE NOT CURRENTLY RECYCLABLE BUT A METAL ROOF WILL LAST LONGER AND IS RECYCLABLE.

Walk-out basements provide great additional space for the home; rec rooms, media rooms and family rooms are all common uses.

> **USE ORIENTED STRAND BOARD (OSB)**
> OSB IS A DENSE FORM OF SHEATHING SIMILAR IN SOME RESPECTS TO PLYWOOD. OSB IS COMPOSED OF CHIPS OF YOUNG GROWTH AND WASTE WOOD, SATURATED WITH AN ADHESIVE AND THEN COMPRESSED TO A FRACTION OF THE ORIGINAL THICKNESS. THE QUANTITY OF THE ADHESIVE IS ACTUALLY VERY LOW AND IT DOES NOT CONTRIBUTE SIGNIFICANTLY TO THE STRENGTH OF THE SHEATHING. RATHER, IT IS THE COMBINED DIRECTIONS OF THE GRAIN IN THE COMPRESSED CHIPS THAT MAKE UP THE COMPOSITE STRENGTH OF THE BOARD. OSB USES NATURAL RESOURCES EFFICIENTLY AND DOES NOT RELY ON OLD GROWTH TIMBER, MAKING IT A GOOD ENVIRONMENTAL CHOICE.

The type and thickness of the insulation affect the R-value of the panels. The most common insulation types are expanded polystyrene, extruded polystyrene and urethane. Expanded polystyrene (EPS) consists of small beads of a styrene foam formed into boards most commonly 4 feet wide by up to 24 feet long (1.2 by 7.3 meters). The thickness of the insulation varies from 3.5 inches to 9.25 inches (9 to 23.5 centimeters) providing R-values from approximately 13 to 35. Extruded polystyrene (XPS) is styrene foam that is forced through molds into panels ranging in sizes from 4 feet wide by up to 24 feet long (1.2 by 7.3 meters). Extruded polystyrene panels are commonly 3.5 inches to 7.5 inches thick (9 to 19 centimeters), offering R-values from approximately 17 to 36. Urethane panels are manufactured by injecting urethane in a liquid form into molds and allowing the urethane to expand and bond to the sheathings. Urethane panels are also usually 4 feet wide and up to 24 feet long (1.2 by 7.3 meters). The range of thicknesses varies from 3.5 inches to 5.5 inches (9 to 14 centimeters), providing R-values from approximately 26 to 40.

All three of the above panel types are available in a variety of sheathing layer configurations. The most common structural panel consists of a layer of oriented strand board (OSB) on each side. In conjunction with a foam core, the panel acts very similarly to a steel beam. When force is applied in a direction perpendicular to the face of the panel, the opposite skin, or sheathing, is placed in tension while the force-side skin is placed in compression. The stiff panel then transfers that load to the frame behind it. SIPs like these are most commonly used in walls to resist and transfer wind loads to the timber frame. They also provide a fastening surface for shingles and clapboards on the exterior wall and for drywall, trim and cabinetry on the interior wall. If SIPs are used in the roof they will provide an interior fastening surface (which is not always necessary) and an exterior fastening surface which can be very useful for attaching the outer layers of the roof system. The use of SIPs in the roof system also allows overhangs because the panels are so strong.

Another common panel type has a layer of OSB on one face and drywall on the other. The most common use for this type is in cases where a drywall ceiling is preferred. The structural properties of this type of panel are much reduced from the type of SIP with OSB on each side, but in most typical snow load applications these panels are perfectly adequate and can span 4 feet (1.2 meters) between rafters or purlins.

> **USING INTERIOR FINISH MATERIALS ON THE OUTSIDE OF THE FRAME**
> IT IS A COMMON PRACTICE TO ATTACH INTERIOR FINISH MATERIALS TO THE OUTSIDE OF THE TIMBER FRAME BEFORE THE SIPS ARE PLACED. THIS IS DONE FOR SEVERAL REASONS. BY ATTACHING DRYWALL TO THE OUTSIDE OF THE FRAME, THE SEAMS CAN BE CONCEALED AND THE AMOUNT OF LABOR REQUIRED TO FINISH VISIBLE SEAMS IS REDUCED. IF THE DRYWALL IS PLACED PERPENDICULAR TO THE ORIENTATION OF THE SIPS, THIS CAN REDUCE THE LIKELIHOOD OF THE DRYWALL JOINTS CRACKING IN THE FUTURE. PRE-FINISHED TONGUE AND GROOVE BOARDS CAN BE ATTACHED TO THE RAFTERS OR PURLINS OF THE CEILING, WHICH REDUCES THE LABOR REQUIRED AND THE AMOUNT OF TIME TO HAVE A FINISHED CEILING.

Ventilation

Ventilation is extremely important because indoor air quality has a significant effect on the health of a home's inhabitants. You cannot count on sufficient introduction of fresh air to the home from opening and closing doors or from the infiltration of air through the wall and roof. Opening windows does provide ventilation, but it also allows heat and humidity to be lost or gained. It is important to implement a system that will remove indoor pollutants from rooms like bathrooms, kitchens and storage rooms—where cleansers and other household chemicals are stored—and bring in fresh outdoor air.

While SIPs greatly improve a building's insulation, homes built with SIPs will be less well-ventilated unless the issue is specifically addressed. In a SIP home that is not being heated and/or cooled with a forced air system, (which will naturally provide ventilation and humidity control) an air exchanger should be installed. The two most common types of air exchangers are heat recovery ventilators (HRVs) and energy recovery ventilators (ERVs).

HRVs use a heat exchanger between the inbound outdoor air and the outbound indoor air, providing fresh air to the home environment and removing stagnant and/or polluted air from it. The heat exchanger passes both air sources past each other through a common heat sink, capturing heat from the indoor exhaust air and warming the cooler outdoor fresh air prior to it going into the home. ERVs are very similar but also transfer the humidity levels of the outbound air to the inbound air, which is important in warmer climates where a cooling cycle is the predominant mode for the building.

Recovery ventilators do not require the same level of ductwork as a forced air system. Average-sized homes need a unit slightly larger than a microwave oven, which is typically suspended from the ceiling in the mechanical room in the basement. If fresh air is not run through a recovery ventilator, the home's heating and/or cooling system must compensate for any losses—so the recovery ventilator provides an energy efficient and effective solution. The benefits of fresh air, improved climate control and energy efficiency greatly outweigh the first costs of a recovery ventilator.

Windows and Doors

A wall with a high R-value such as R-26 will typically have sections replaced with windows and doors that have substantially lower R-values. Although doors and windows are obviously important for light, ventilation and access to the home, their selection and placement are also important for energy efficiency.

There are a few options to consider when choosing windows, many related to the type of glass used. Options include double pane, triple pane, low-emissivity (or low-e, a coating that helps keep heat within the home), tinted and reflective. To save significantly on heating costs, it is strongly recommended that you choose durable windows that insulate well. I recommend double or triple pane glass that is low-e coated. Aluminum-clad wood frame windows are a good choice because wood can be replenished easily and aluminum is durable and can be recycled.

Doors should be of high quality materials and seal the home well from the elements. Solid doors are most commonly

---

SIPS AND VENTING PANELS

SIPS PROVIDE BOTH THE INSULATION AND THE VAPOR BARRIER IN THE WALL AND ROOF. HOWEVER, IT IS IMPRACTICAL TO REQUIRE PERFECTION FROM ANY MATERIAL OR ITS INSTALLATION. EXTREME CONDITIONS (SUCH AS HIGH INDOOR HUMIDITY OR SEVERE STORMS) OR EVEN MINOR IMPERFECTIONS IN SEALING THE PANEL JOINTS WILL ALLOW MOISTURE TO BECOME TRAPPED BETWEEN THE EXTERIOR BUILDING MATERIALS AND THE PANELS. IF MOISTURE REMAINS THERE, EITHER BEHIND AN AIR BARRIER OR TRAPPED BETWEEN MATERIALS, BOTH THE MATERIAL AND THE OSB FACE OF THE PANEL WILL SUSCEPTIBLE TO MOLD, ROT AND DETERIORATION. IT IS THEREFORE STRONGLY RECOMMENDED THAT A VENTING PRODUCT OR STRAPPING BE INSTALLED WITH THE WALLS AND ROOF TO FACILITATE AIR MOVEMENT BETWEEN THE DIFFERENT LAYERS. DISCUSS THE OPTIONS WITH YOUR HOME DESIGNER AND BUILDER.

SELECTION OF LOW-IMPACT MATERIALS

THE BUILDING INDUSTRY IS A MAJOR GLOBAL CONSUMER OF BOTH RAW MATERIALS AND FOSSIL FUELS. ANY REDUCTION THAT CAN BE MADE IN OUR CONSUMPTION OF THESE THINGS WILL PROVIDE US AND FUTURE GENERATIONS A HEALTHIER WORLD TO LIVE IN. YOU CAN LOOK FOR MATERIALS THAT HAVE LESS OF AN IMPACT ON OUR NATURAL RESOURCES. SOME EXAMPLES INCLUDE CONCRETE THAT SUBSTITUTES FLY-ASH (A BY-PRODUCT OF BURNING COAL) INSTEAD OF VIRGIN PORTLAND CEMENT; CARPETS, CARPET PADS, INSULATION AND EVEN FLOOR TILES THAT ARE HIGH IN RECYCLED CONTENT; AND FLOORING MATERIALS MANUFACTURED FROM HIGHLY RENEWABLE RESOURCES SUCH AS BAMBOO OR WHEAT BY-PRODUCTS. ALL OF THESE CHOICES MAY APPEAR SMALL BUT WHEN TAKEN TOGETHER THEY CAN HAVE A SIGNIFICANT AND POSITIVE IMPACT.

wood-veneered, where the hidden core is made of smaller pieces of wood. These smaller pieces make the door stronger and greatly reduce the probability of the door warping or shrinking, which would compromise its seal to the outdoors. Glazed windows for glass doors are available in the same high quality materials as other windows and have reasonably good insulation properties.

LIGHTING AND POWER

The lighting and power requirements of your home may range from typical outlets and light fixtures to the specialty requirements of large home theaters or luxury home spas with lap pools, saunas or hot tubs; the latter of which will affect the overall size of the electrical service brought into the home. It is best to identify and address lighting and power requirements early in the design so that they can be planned for accordingly. In some cases it is appropriate to hire specialty consultants in order to avoid re-design in later stages of the project when rewiring changes will be more costly.

Typical homes have duplex outlets along the perimeter of all walls and along countertops in kitchens. Dedicated outlets must be provided for common household appliances and ground fault-protected outlets are required in bathrooms. Electrical codes and convenience will dictate the minimum number of outlets and how many dedicated circuits will be needed.

In this phase of design, discussions about lighting should address the qualities and quantities desired for certain spaces and functions. Lighting has a huge impact on the perception of space. The choice of fluorescent or incandescent will affect the perceived warmth of a space because each emits a different spectrum of light.

The lighting of a space should be appropriate for that space's intended uses. If a space has a single primary function, such as the dining room, then consider multiple fixtures that provide an even, warm and non-glaring light. A loft that will serve double-duty as an office and play area for children might need specific task lighting sufficient for reading, writing and computer work, while the space as a whole may need to have an even, bright white light so that children can play.

REVIEW THE PLANS AT LEAST ONE MORE TIME BEFORE BUILDING BEGINS TO IDENTIFY ANY ADDITIONAL AREAS WHERE YOU FEEL ELECTRICAL ACCESS WOULD BE USEFUL AND OUTLETS SHOULD BE INSTALLED.

The trellises provide shade from the high summer sun but admit light when the sun is lower in the winter months.

PROJECT SCOPE AND RESPONSIBILITIES

Everything works better when the project scope and responsibilities are clearly defined in this first stage of the design process.

Start by identifying the project scope, including the sum total of products, services, requirements and features of the project and their relation to its involved parties: client (you), architect, designer, contractor and consultants. All parties should clearly understand who is responsible for each service or product.

Define what the final product will be as well as specific beginning and end points of construction. If construction is to be phased, define the phases. Identifying the project's scope at this stage allows you to monitor "project creep", a term used for the issues that arise as additional items are added to the project during the process while ignoring budget and schedule! Examples of project creep include deciding to finish the basement when it was originally meant to be left unfinished; or adding a bathroom or closet at the last minute.

PROJECT SCHEDULES

Project scheduling is principally done by the contractor. Scheduling consists of identifying the primary construction elements or work structures. These are broken out by trade or building system, such as plumbing or roofing, and then assigned start and completion dates. These work breakdowns will have estimates for labor and materials that coincide with each line item, allowing the materials to be ordered and for material and labor to be appropriately budgeted. It is important to remember that a schedule is an estimate, requires the consensus of all those involved, and is always subject to the wonderful inconsistencies of chance and nature.

Schedules are often represented graphically, using Gantt charts. A Gantt chart illustrates the phases and activities of the project by each work item, usually by week. These charts are used as schedule management tools and are helpful in visualizing the dates at which each item is expected to be complete, and the dependencies for each phase.

PROJECT BUDGET AND CONTINGENCIES

The project budget can be set by the maximum amount allowed to be invested into the project according to bank support, or set by a personal spending limit. Budgeting is difficult at this stage because of the sheer number of undefined variables—for example, will the countertops in the kitchen be quarried granite or plastic laminate? When these items are not defined, an allowance per square foot, lineal foot or by room can sometimes be fairly accurately estimated. Contractors often have examples of the types or levels of finish that fall into these allowance brackets that will help you arrive at these estimates.

In this early stage, budgeting can be done in a very rough form by using a total square foot cost of construction multiplied by the anticipated size of the home. The square foot cost should be based on comparative regional construction costs in conjunction with comparative costs for construction type and level of finish. For example, the contractor should have construction cost figures for a home that is similar to the anticipated project in size, complexity, interior and exterior materials and fixtures. For the sake of this example, let's use a cost of $200 per square foot. If the initial size of the home is estimated to be between 2000 and 2200 square feet, then you can estimate that a budget of $400,000-$440,000 will be required. The numbers you arrive at using this method are rough estimates, but can provide very good guidelines in the preliminary stage of design and keep you from getting in over your head farther along in the process.

Another important aspect of budgeting is to make allowances for the unexpected. Construction delays can be caused by weather, mistakes, or project creep. All of these are common realities. These occurrences can be planned for by allowing a contingency fund, typically 5-10% of the total anticipated project cost. For the above example, $20,000-$40,000 should be set aside to comfortably accommodate the unexpected.

chapter four
04

REFINING THE DESIGN

THE FOCUS OF THE NEXT FEW PHASES OF DESIGNING YOUR HOME IS THE PROGRESSIVE REFINEMENT OF YOUR IDEAS AND DECISIONS MADE IN THE PRE-DESIGN PHASE. THIS INVOLVES THE PRODUCTION OF SCHEMATIC DESIGNS AS WELL AS FURTHER DESIGN DEVELOPMENT. THE ARCHITECT WILL START BY PRODUCING SCHEMATICS FOR DISCUSSION AND REVISION. ONCE THESE ARE APPROVED, THE DESIGNER OR ARCHITECT WILL FINESSE THE DETAILS IN THE DESIGN DEVELOPMENT PHASE.

SCHEMATIC DESIGNS

Schematic design drawings are prepared by the architect with the purpose of conveying a design concept to you. There may be one or several sets of documents, depending on how close their initial design concepts align with your vision. In these documents, you will find that basic spaces and their relationships to one another and "massing" (which is the process of defining the physical volumes and forms of different parts of the home) are of primary importance. Note that not all of the building systems, rooms or aesthetic concepts will be represented at this stage.

Due to the nature of schematic design drawings, some explanation from the architect or designer will be necessary to adequately convey what is not illustrated in the drawings. When you are presented with a visual representation, you might find that it works as a catalyst for further exploration of the design vision, re-examination of the initial concepts or even investigation of new avenues that the design might take. Document initial reactions and questions from both parties, as this will help direct the next phase.

While the main function of schematic design drawings is for you and the architect to convey, discuss and respond to the design concept, you can also give early schematic drawings to the contractor for cost estimating purposes and preliminary schedule investigation. Consultants like structural engineers and civil engineers can also benefit from these drawings for preliminary engineering and investigation of foundation systems, or preliminary grading, site access studies and utilities layout planning.

NURTURING IDEAS

The schematic design phase is a good time to challenge your initial vision, outline and design concept. Revisit the organization and adjacencies of spaces, and note the advantages and disadvantages of differing solutions, based on the most recent, updated information. These interactions between you and the architect can be some of the most important. Not only is your vision addressed in the context of actual design, but also in its relation to budget, materials, construction methods and technologies.

FIRST STEPS

When an architect sits down to come up with the first schematic design, there is nothing but text, words, ideas, photographs and a site to consider. How does a design arise from this? This is where your hard work during earlier planning and pre-design stages comes into play. Your notebook, scrapbook of images and conversations will help inform the next step. For example, perhaps you have decided that a distinctly western style of home is what you want. If you conveyed this effectively to the architect, it will be the seed of the architect's concept for their schematic drawings. Alternatively, seed ideas might come from visits to the site that you and the architect do together; perhaps a sloping site with great southern exposure and a relatively flat area inspires the idea of nesting the home on the northern side of the flat area and letting the home encircle that area so that it might become the center of outdoor activities.

Once the architect has developed some seed ideas with you, they can begin to explore the "massing" of the structure, or give it rough conceptual size and form in conjunction with addressing site and home adjacency requirements. Adjacency diagrams often take the form of bubble diagrams in which circles of different sizes are drawn to represent rooms or spaces and arrows are used to indicate important connections.

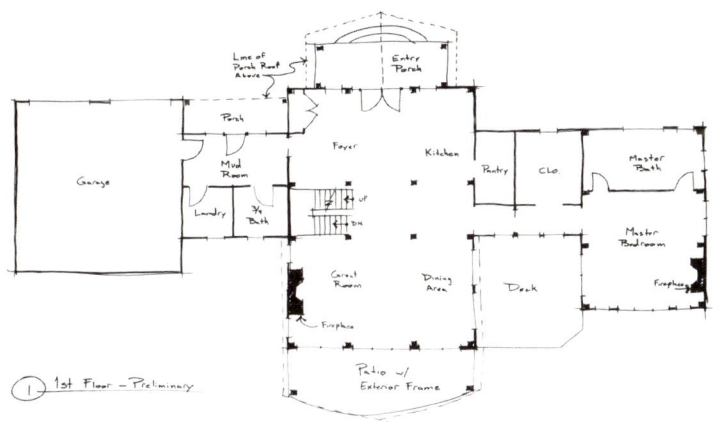

A sketched plan can be developed directly from the bubble diagram and helps to clarify the sizes of spaces and the flow of traffic within the home.

Massing studies are simply ways to assist with visualizing the structure (your home) in a physical form. Massing studies can take the form of hand drawn sketches, rough-built cardboard models or three-dimensional masses drawn with a computer program. All of these methods are appropriate and can be used independently or together. It can be very useful to view these alternative forms of illustration.

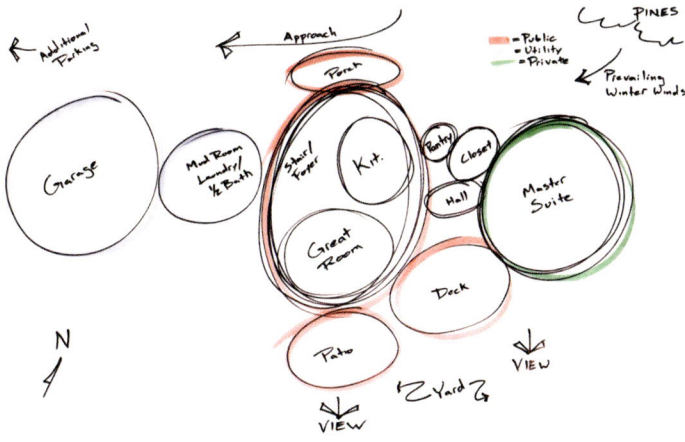

A bubble diagram illustrates important functional adjacencies and the appropriate location of the spaces on the building site.

Computer massing models help convey such things as the scale and proportions of a home.

Here the process of massing begins to respond to your vision and the site considerations identified in the pre-design phase. These items can appear in sketch form on the drawings as arrows with labels that identify specific characteristics and the design responses that are meant to address them. Although some of these sketches may not make it into final presentation drawings, they are an important part of the process.

REFINING THE DESIGN | 103

During this phase, it's important to check the overall square footage. Use rough calculations based on the massing models. Depending on the size of the home, these numbers should be within 15% of the goal square footage—if not, problems will almost certainly arise later in the process. Be aware that balancing the outline goals and the square footage is challenging because of the lack of exactitude in the drawings at this stage. However, anything that supports meeting the project and budget goals must remain of prime importance throughout the process.

Schematic drawings tend to start out as rough sketches. Circles within larger circles may begin to define rooms or even spaces within a room. There may be only one drawing that contains several iterations drawn atop each other in multiple colors; or these iterative sketches might develop into a half-inch thick stack of tracing paper, each with a different version of the layout.

Keep in mind that these schematic drawings can go through several iterations at this stage. Some will reflect partial ideas that only work independent of the site; others might not work with factors like the site topography; and yet still others will have nothing explicitly wrong with them—they just "don't feel right". However, many drawings will feel like they do reflect your vision. These design ideas and elements will be taken to the next step. They represent the pieces that, if nurtured, will evolve most naturally into the right home for you.

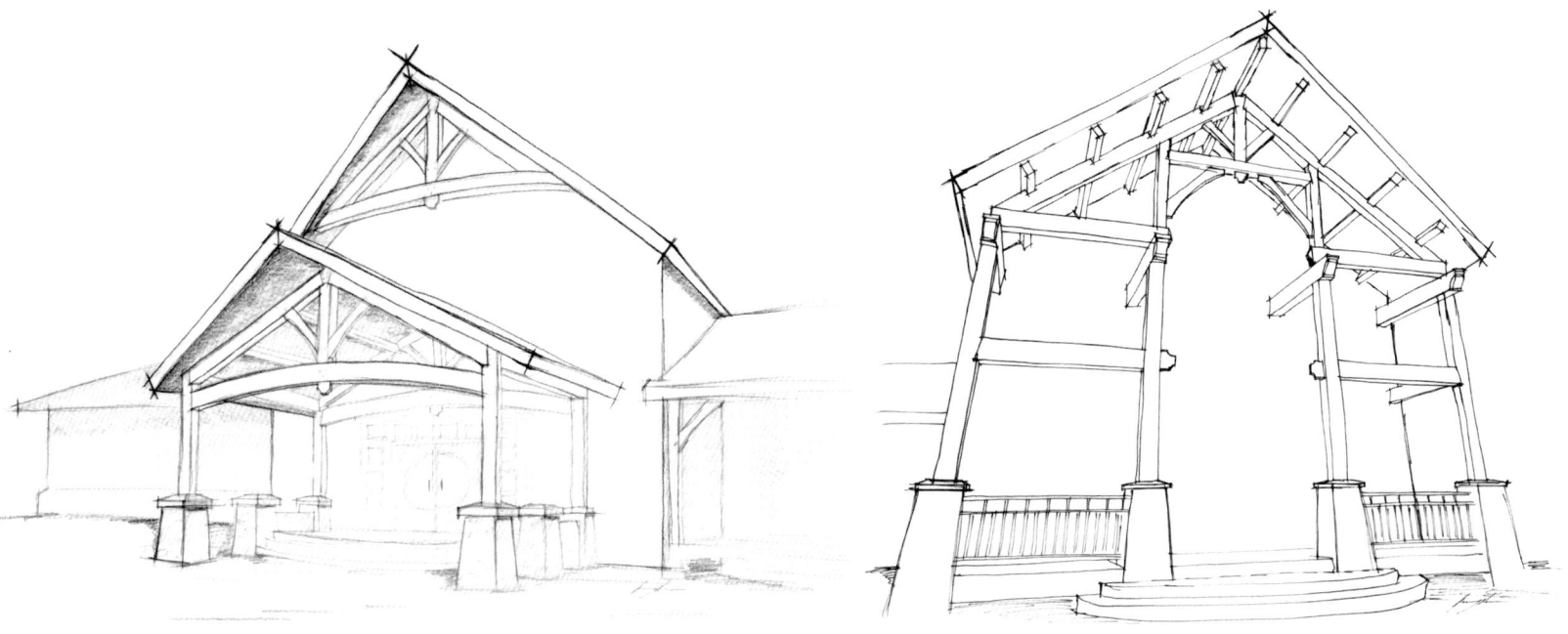

Detailed character sketches of a home's front and back entrance can help establish the character of a home and help develop ideas for the timber frame structure.

INTERMEDIATE STEPS

If the style of the home was defined in your vision, it will come into play here. For example, perhaps you expressed the desire for the home to appear nestled within the site as opposed to perched upon it. As a result, the architect may suggest that the structure or massing of the house be broken into smaller appropriate sizes that relate to site-specific characteristics. It is important at this stage for the architect to pay very close attention to your reactions and for both parties to communicate as clearly as possible. Both parties should be asking questions and clarifying answers, as well as stating likes and dislikes.

Within this stage the defining characteristics may begin to evolve into small sketches of spaces, layout diagrams for specific rooms or even labels or lists on the drawings themselves. Toward the end of this stage, design drawings will begin to look even more like recognizable forms of a home; roof planes, terraced areas, porches and balconies will start to appear in the sketches and models.

This is the time that the initial ideas for timber framing of the home will be considered. Issues include whether the home will be partially or entirely timber framed, and how to address the implications introduced by such issues as frame types or post layouts in relation to specific room design ideas.

Preliminary thought should be given to site utilities and systems for power, water and waste, with attention to the site constraints. The home's placement in conjunction with the drilling of a well for potable water, required placement of a septic system or property setbacks may restrict the footprint of the home or even require reassessment of the placement of the home on the site.

Building, zoning and specialty regulations are also best reviewed at this stage, when the design, massing and layouts are still rough in design and can be manipulated easily.

FINAL STEPS

The final stage of schematic design drawings should be well-developed and address massing; room and space layout; locations of stairs; interior circulation; and timber frame extents, locations, style and layout. Window and door locations as well as primary views and access as they relate to site should be indicated. These drawings will take the form of two-dimensional plans of the primary floors of the home, two-dimensional elevations of the exterior; cross-sections through the most important areas of the home and possibly sketched or computer-rendered studies of the massing of the home. Types of construction should be delineated, indicating the parts that will be timber framed and the parts that are to be conventionally built.

From here the design will proceed into the design development phase, where detail becomes the focus, now that the larger, more encompassing foundation has been laid.

TIMBER FRAME PLANNING

As mentioned earlier, at this stage the timber frame plan is just beginning to develop in unison with the design. You might have had specific ideas since the beginning, like a cathedral-ceiling great room looking out on a meadow or mountain range. This is often the case if something specific sparked your vision, like a friend's timber frame home or a photograph. In these instances, you may have already decided upon the frame type itself. If the timber frame design itself was integral to your vision, then the architect will probably have already created some sketches, and might even develop some three-dimensional models from a CAD (computer aided design) program for you.

The process of massing sometimes lends itself to a particular frame type because of a physical constraint like a large span or the preference for an open floor plan. In other words, the spaces, size and layout will determine the span and thus help determine the

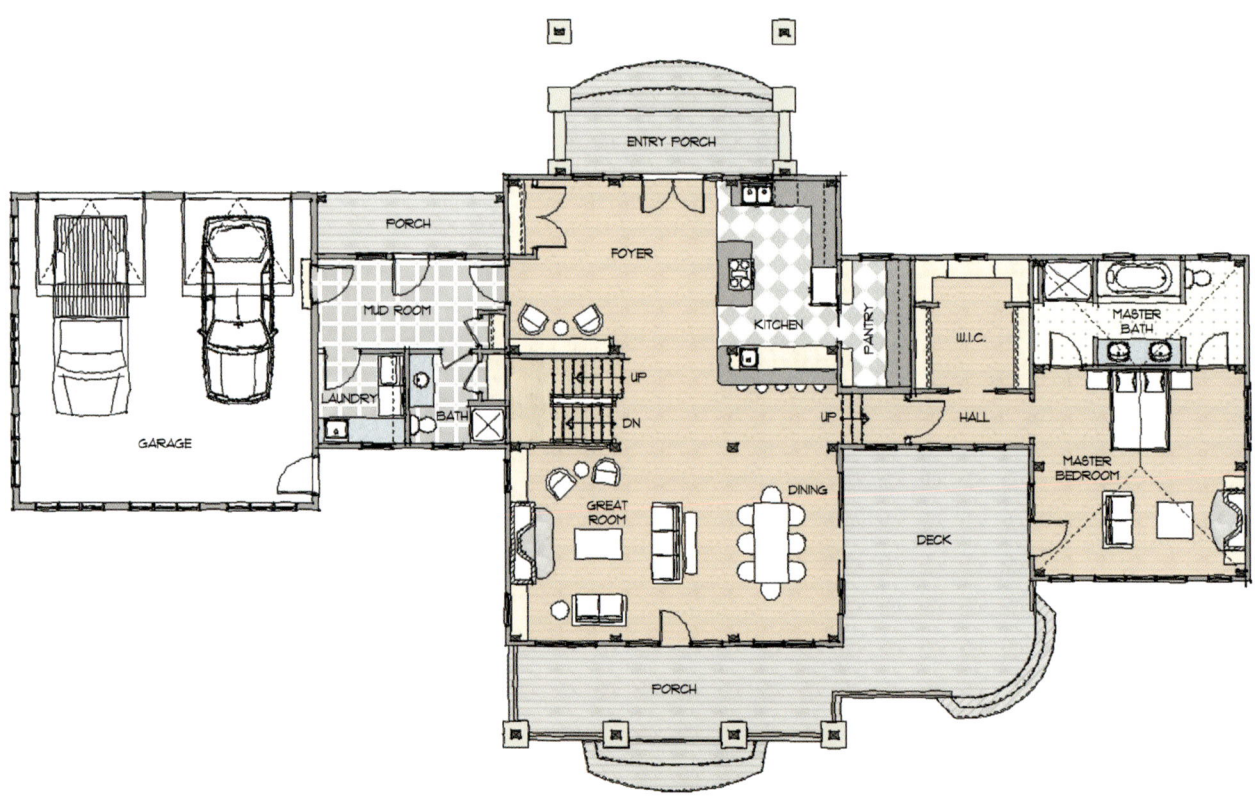

A two-dimensional computer-rendered plan with furnishings and fixtures included can help you to better understand the intended use of space in the plan.

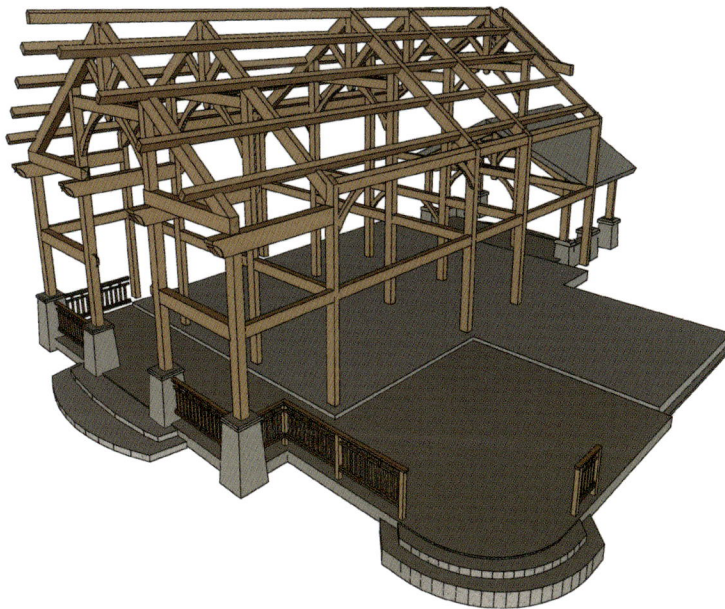

A computer-rendered model of the frame (in this case, common purlin) can give you a sense of how the frame relates to the form of the building.

type of frame. The reverse can also occur, where your desire for a particular style of timber frame will determine the overall size and structure of a room.

It really can go either way; the key is to make sure that, as always, these room-specific design decisions are made with consideration of the effects on the home design as a whole.

DESIGN DEVELOPMENT

Until this point, the process has addressed the project as a whole, focusing on the breadth of the design, attending to the issues of layout, adjacencies, massing and scale on a macro level. The previous phases of design have given focus to the vision as it relates to the home perceived from a distance, both figuratively and realistically.

Throughout the design development phase, attention is given to individual aspects: each space or room, window and doors, material selections and locations, building systems and structure. Design development concentrates on finalizing ideas and refining spaces and technical aspects of the home. Architects and designers are able to convey design concepts to you in much greater detail, by illustrating the functions of spaces and adjacencies and giving further detail to what these spaces will look and feel like. The drawings will contain detailed plans and elevations including more sections throughout the building, and a refined site layout with details relevant to their construction. During this process, reactions and questions should be documented to confirm the process is proceeding in accordance with the guiding principles of your vision and outline.

The drawings during this phase can begin to take the appearance of formal construction documents; CAD drawings give the appearance of completeness and accuracy. These drawings will aid the builder in developing a more accurate estimate but they should not be considered formal construction documents. It's important to remember that this stage is still part of an evolutionary process. Dimensions will change as walls, rooms and other aspects shift in their nature and dimensions. Developing any system or space, as is required at this stage, will certainly affect others. The timber frame itself can appear finalized, set in its dimensions and timber sizes, but be sure to leave room for changes.

The real purpose of design development drawings is to define and illustrate as many elements of the home as possible. After the design development phase, any changes in the project could affect budget and schedule drastically. Therefore, agreement about the scope and design of the home at the end of this phase is critical.

FIRST STEPS

The first steps of design development typically involve working with a compilation of the best ideas from the schematic design phase and integrating them into a solution. This solution will then be examined from the vantages of scale, proportion, aesthetics and the overall plan, as well as the integration of the building systems (heating and cooling, fuel, utility locations and required materials).

Decisions about some of these things might involve the builder to a great extent. For example, you may want them to participate in the research and selection of materials and fixtures. Materials research is often greatly assisted by the builder's familiarity with

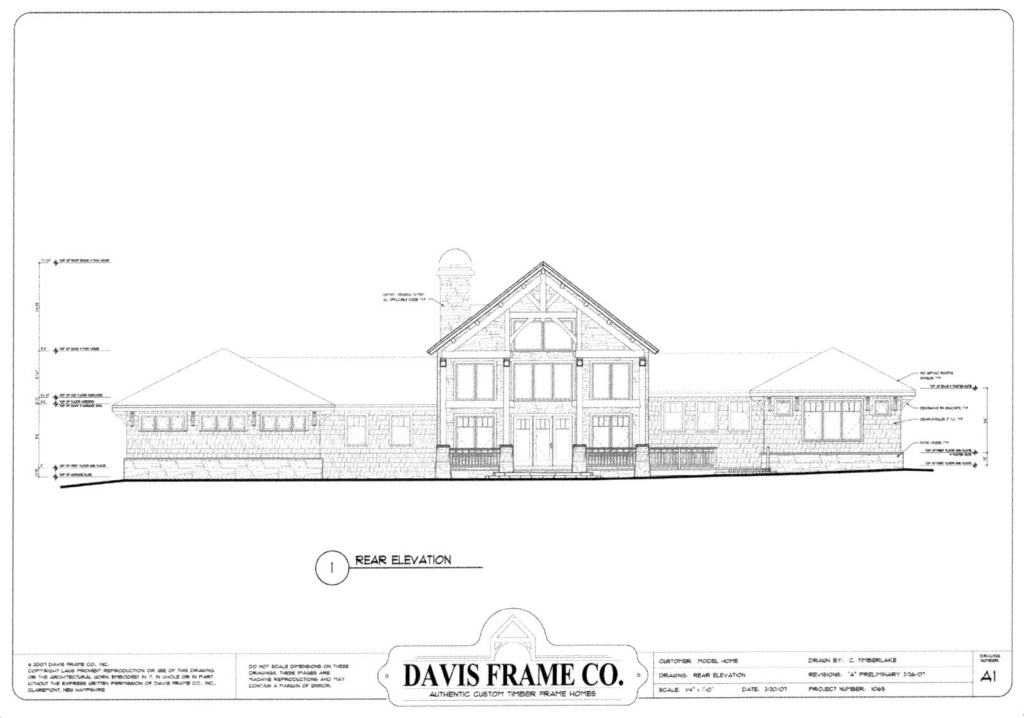

A construction elevation depicts important elements such as frame heights, building heights, materials and windows.

both the process and local availability. The builder may also have previous homes photographed or open for visits that can assist you in making decisions about materials or finishes.

At this point, the timber frame and other structural system designs are well underway. The framing styles and wood species should already be selected because the species selection can affect the structural characteristics of the frame dramatically. By the end of the design development stage, most major timber elements should be sized and the frame design close to completion.

The drawings in the early stages of this phase can include enlarged floor plans of rooms or important pieces of the plan. All elevations should be drawn and enlarged elevation details developed to convey unique or important characteristics. The timber frame can be designed in a CAD program, which allows isometrics or different perspectives of the frame and home to be viewed. Sketches and studies of details, both interior and exterior, should begin here for exploration and confirmation from the owner.

INTERMEDIATE STEPS

As the drawings become more detailed and more of the character and quality of different spaces emerges, your feedback becomes even more important. Rooms will begin to have clear dimensions and ceiling height defined. It's critical to review the sizes of these spaces and assess if they will indeed function properly for the activity that they are intended for. Looking closely at the plan and section of a great room, for example, will ensure that the intended amount of furniture will fit into the space and if it meets your anticipated use requirements of the room. The section drawings will aid in understanding the height of the spaces, delineating both higher ceiling spaces where more people can gather in comfort, as well as lower ceiling areas where more intimate and smaller gatherings can occur. These drawings must be studied and discussed between you and the architect together to fully understand the quality of these different spaces.

A heavy scissor truss with a bolted steel connection complements the character of this home.

FINAL STEPS

In the final stages of design development, all regulatory guidelines, including zoning regulations, building code and specialty regulations should be in the process of being addressed. The building department or design review boards should also be consulted at this stage, which may require preliminary submissions for review and comment. Input from the authorities is important at this point because changes to the project scope or program could have negative schedule and/or budgetary impacts if they are not addressed now.

All major building systems should be addressed and identified in this set of drawings. Consultants like structural engineers or kitchen designers should have their design development drawings back to the architect at this point as well, so that integration of their designs can occur.

The drawings at the end of design development should be hard-lined by hand or computer-drafted. They should include overall dimensions for the building and its contingent parts; rooms and their dimensions should be labeled; all fixtures for baths and kitchens as well as major appliances indicated. The plans can also indicate materials either directly or in schedules. Another option is a specifications list prepared either by the designer or the builder. Elevations should indicate dimensions of ridge heights and floor heights. Illustrations of cross-sections throughout the building should indicate ceiling heights and depict areas of transition between the stories of the home such as stairs or lofted spaces. Details of specific interior or exterior characteristics can be depicted to convey the character of design elements, like exterior timber brackets or an interior entryway arch, for example. Building isometrics, frame isometrics or perspective may also be prepared at this stage to convey the exterior or interior characteristics of the design.

These drawings should convey to the owner and to the builder all of the possible elements of the design. To be blunt, this is quite a feat. The range of issues is vast, addressing issues such as "what is the most appropriate flooring material choice in the entryway" to, "at what time will the sun stream into this room, because this is where I will spend my Sunday mornings reading?" The drawings should answer these questions, and at times may also provoke more questions. However the end of this phase should represent near-completion and agreement between you, the builder and architect of the design as a whole.

A computer-rendered model represents one of the final stages of design and is meant to convey the scale, colors and overall aesthetic of the home.

chapter five

CONSTRUCTION DOCUMENTS

Construction documents are intended to convey the scope of design to the builder so that the home can be priced, scheduled and built accurately. Construction documents are also used by the building code official to verify that a project is meeting all health, safety and welfare requirements. It is in this phase that details and systems are completed and coordinated within the whole of the design.

This phase takes the design as developed and finalized during the design development stage and records it for construction. All major design issues have been determined by now. Documentation should include plenty of detail, and all drawings should be refined as each construction issue is addressed. The emphasis in this phase is to achieve a level of clarity and completeness in the design drawings that allows the project to be built efficiently.

Drawings before this point have been focused on conveying the design, its character and aesthetic to the owner to make sure that the vision and outline will be met, and to provoke discussions in an effort to identify any shortcomings. This final evolution of the drawings addresses the technical issues of constructability. The range of drawings produced will vary depending upon the complexity and size of the design. There will be architectural drawings, timber frame drawings and panel drawings. There might be mechanical systems drawings, electrical layouts, plumbing diagrams or specialty drawings for cabinetry and millwork. However many drawings and whatever type, they will convey to the builder what is to be built. These drawings tell the story of the construction of the home.

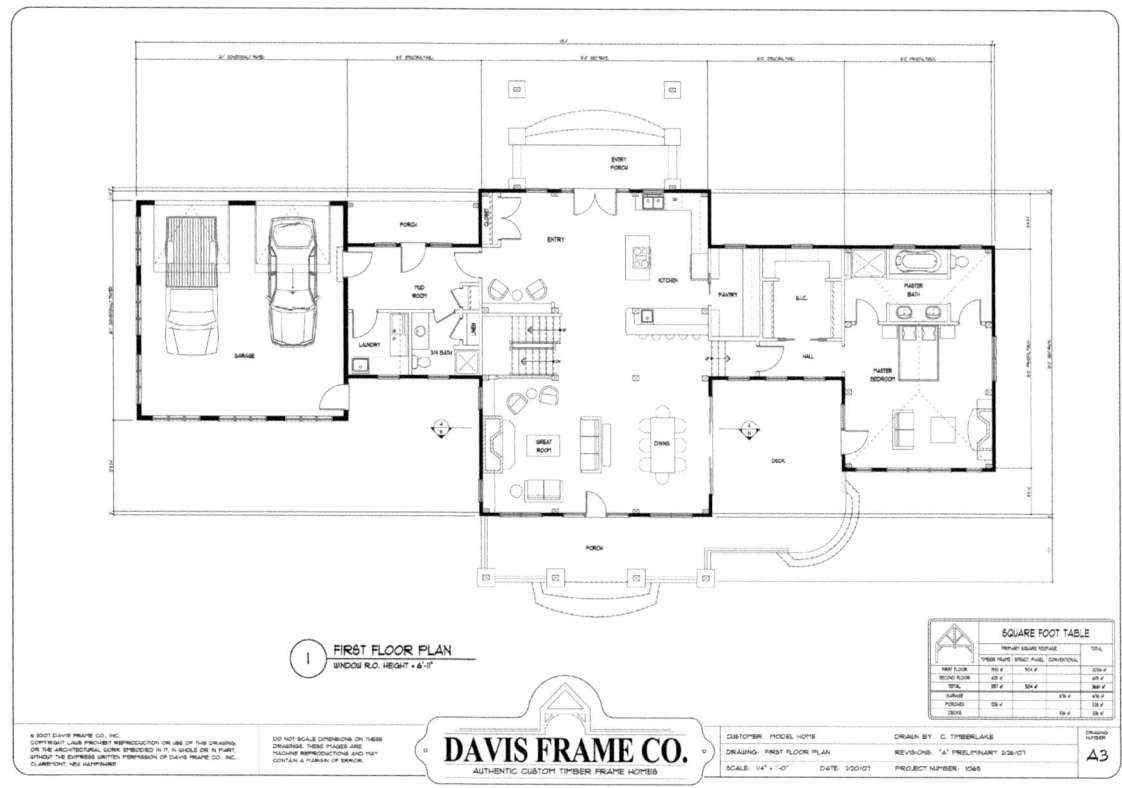

Floor plans communicate critical information such as dimensions for walls, doors, windows and stairs. They also indicate materials and finishes, and reference other detailed construction drawings that provide more information.

ARCHITECTURAL DRAWINGS

Architectural drawings convey information in a variety of ways. Each of these drawings depicts either a particular part of the building specifically or the entire building in a specific manner. All architects or engineers will do these drawings slightly differently. The following sections describe the most common documents for a home.

COVER SHEET

The cover sheet often provides a rendering or elevation of the home. It will also contain project-specific information such as your name, site address, architect and builder names, applicable building code information and square footages.

FLOOR PLANS

Floor plans contain a tremendous amount of information and are the source for the majority of cross-references in the construction set. Cross-references are tags that reference another drawing or sheet within the construction set. These references are crucial in understanding the building because they direct the builder to information that is not indicated in the floor plan.

Floor plans are basically a view of the home as if it were sliced several feet above the floor height, looking down. This slice or section of the home will pass through most doors, windows and walls. It will also indicate other interior items such as counters, stair openings and plumbing fixtures and their dimensions. Other annotations should include reference floor elevations, elevation references, building and wall section

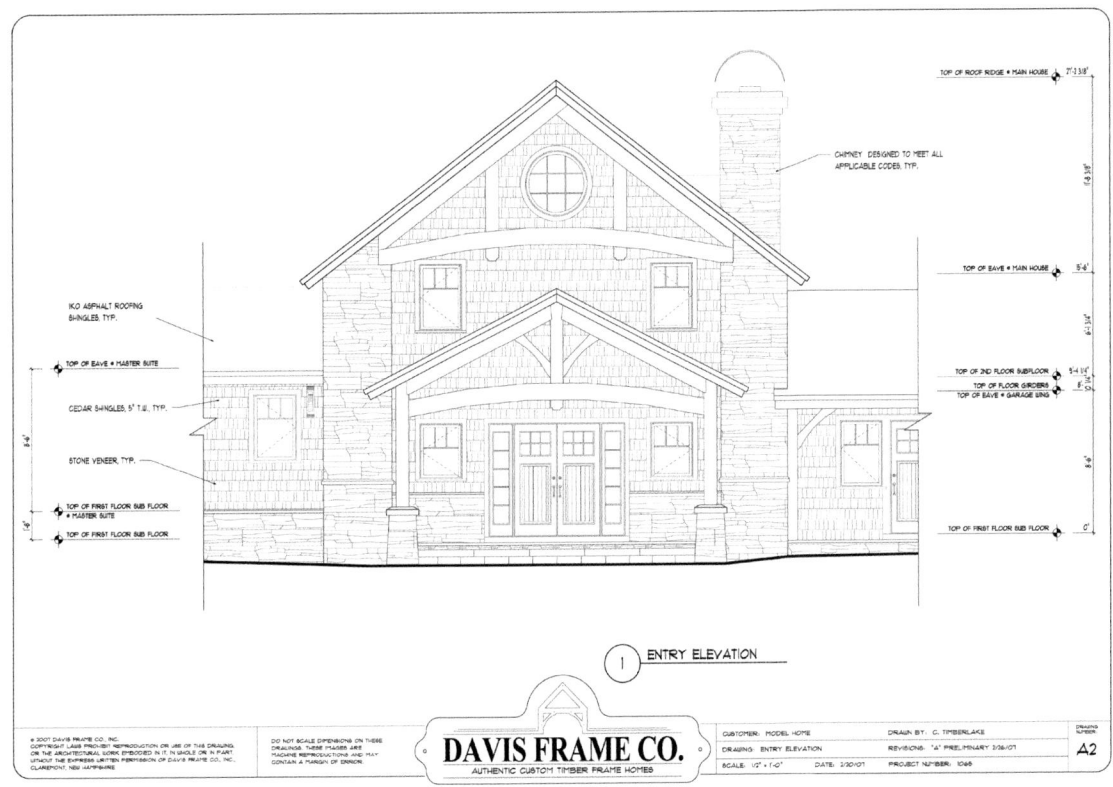

Elevations aid in identifying finish materials such as clapboards, shingles and trim, as well as colors.

references, detail references, room names, and window and door tags. Ceiling conditions can also be indicated if the construction does not require a reflected ceiling plan; vaulted areas, soffits or arched openings are examples.

ELEVATIONS

Elevations are views of the home taken from the exterior, typically perpendicular to the primary facades of the home. The elevations will indicate exterior materials such as siding and roof material, as well as give dimensions in the vertical plane for floor heights or overall building height. Roof pitches and exterior elements such as balconies or decks should also be shown and annotated.

BUILDING SECTIONS

Building sections are also like slices through the building, only in this case they run through the building vertically. These sections show the architectural and structural relationships; how the walls, floors, ceilings and roof work together structurally; and show relationships between the floors of the building, perhaps by passing through stair openings or lofted spaces that communicate with the space below. Building sections can also indicate mechanical information such as ductwork. The building section is another reference drawing that refers the builder to other drawings, such as wall sections or details, where further information can be found, typically at a larger scale.

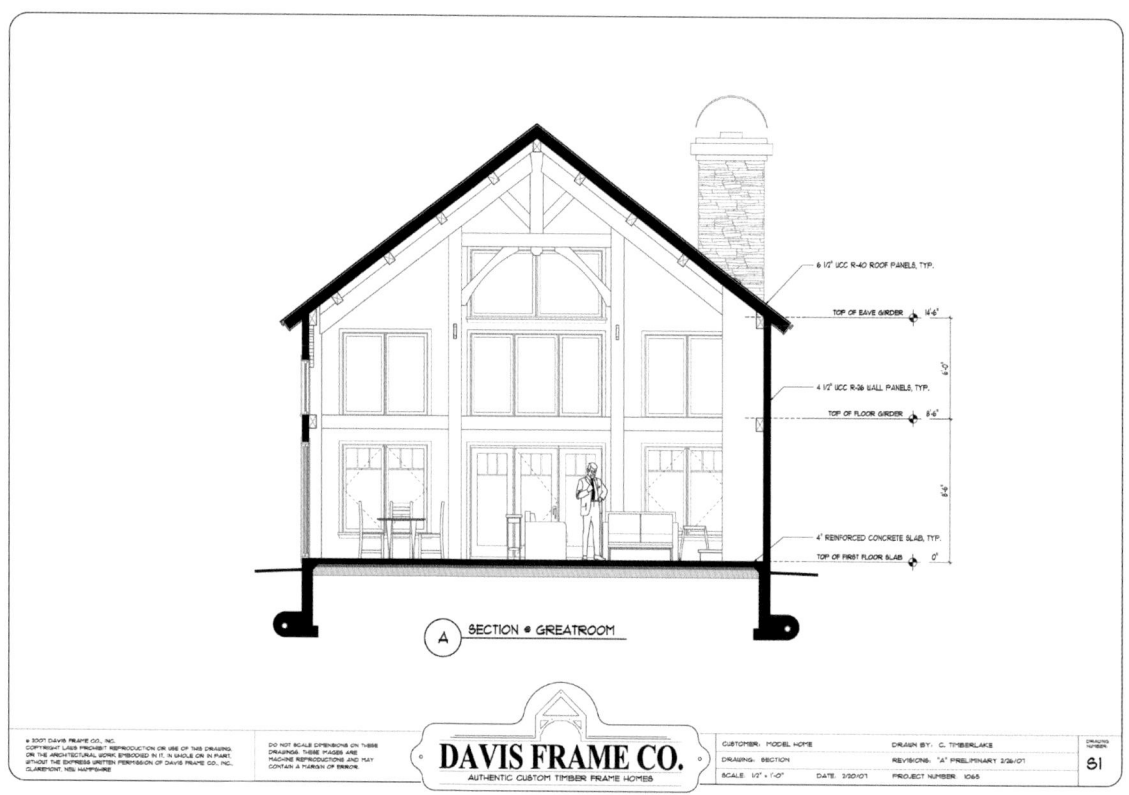

Building sections illustrate interior volumes and indicate important building systems within the floor and wall cavities.

FOUNDATION PLANS

Foundation plans are similar in their development to the floor plans in that they represent a horizontal slice, looking at the foundation system or systems for the home. Foundation plans will indicate foundation types, materials, footings and steel reinforcing. They will be dimensioned in a similar fashion to the floor plans.

SECTIONS AND DETAILS

Sections and details are typically the largest-scale drawings in the set. They are enlargements, in horizontal or vertical cross-section, of specific instances of construction within the building, such as an eave detail at the roof edge. These details will commonly include pertinent dimensions for construction as well as information about additional materials that may not be recognized at smaller scale drawings, such as flashings and fasteners. The primary function of sections and details is to illustrate common and unique construction types within the building. For example, a wall section that illustrates the construction style of 90% of the walls in the house might be indicated as a typical wall section, while the remaining 10% of the walls will require two or more section drawings to illustrate how they are to be built differently from the typical walls.

SCHEDULES

Schedules are developed to correlate extensive information to a tag indication in the drawings; for example, a schedule can contain the text information that corresponds to the reference tags in a floor plan drawing. This system reduces the clutter that can occur from placing large amounts of text in a drawing. The

most typical schedules are door, window and finish schedule. Door and window schedules will commonly indicate the location of the door or window, manufacturer, material, color, unit size and rough opening size. Finish schedules will indicate the room, floor finish, wall finish, ceiling finish and in some cases baseboard, wainscoting or other trim materials and colors.

FRAME DRAWINGS

Frame drawings will consist of the frame dimensions and references needed for the erection of the frame. Drawings are produced for the entire house, so they refer to both timber framed portions and conventionally framed portions. This plan will commonly reference wall views of the frame, such as a particular bent or gable wall, for example, and then illustrate that wall in an elevation view. Roof framing will also be indicated in a plan or horizontal view, commonly referred to as rafter plans, which will illustrate and provide dimensions for all principal purlins or rafters and all common purlins or rafters. Frame drawings will correlate to a numbering system used on the timbers, a critical aspect of constructing the timber frame in the field.

PANEL DRAWINGS

Panel drawings are critical during construction because they illustrate and correlate to a panel numbering system. In the case of SIPs, panel drawings are created for wall and roof panel applications. Panel drawings indicate the layout of panels on the structure; each wall has its own drawing. Most roofs are dealt with in one drawing.

AFTERWORD

"IF YOU INVEST IN BEAUTY, IT WILL REMAIN WITH YOU ALL THE DAYS OF YOUR LIFE." – **FRANK LLOYD WRIGHT**

Designing and building your custom home can sometimes feel like traveling down a bumpy, winding road, with stops and starts along the way, and there may even be times when it will feel like you are traveling in circles. Speaking from experience, having been through this process many times before, I can assure you that the journey will be worth it.

Building and residing in a timber frame home ties you to an ancient craft and venerable tradition that connects you to the earliest American settlers, centuries of European construction and the Egypt of the pharaohs. Building methods may have changed with structural insulated panels (SIPs) replacing wattle and daub, and cranes replacing derricks and horse teams, but the structural principles, forms and the way that timbers are assembled—which lie at the core of this tradition—remain the same.

I'm sure my enthusiasm for the SIPs innovation has been made clear throughout this book. Other than the fact that they are not traditional, I just can't see a down side to choosing them. Using SIPs means that your house will last longer and use less fuel for heating and cooling, which delivers the twin benefits of reducing your consumption of natural resources and saving you money. We can all make choices that are more mindful of the planet that sustains us. What can come as a surprise to many is that the more environmentally friendly option is often the most economical by far, especially when considered over the long term.

A final word of advice: choose carefully the people who you work with. Designing and building a home involves a changing cast of characters, but the core team consisting of your designer and architect, the builder, and your key contractors should be selected based on their reputation for quality and the passion they have for their work. The design process can be a long and occasionally challenging one; and it will be immeasurably easier if you are working with designers and builders whom you trust and are comfortable communicating with. You may not be an expert in designing homes, but you are an expert in what you need and like, so remember that your input throughout the design process is critical for the result to be a home you love.

Timber frames lend a unique quality and character to a home by formalizing continuity and order; representing stability, longevity and comfort; and exemplifying simplicity, tradition and beauty. I wish you the best of luck in your designing adventure and hope this book helps make the journey to your ideal timber frame home more sure-footed and enjoyable.

GLOSSARY

aisled frame – a style of timber frame that can be used to create wider buildings. The aisled frame uses internal posts that divide the floor plan into bays, including aisles along the eave walls and a central bay or nave.

awning window – a type of window that is operated with a crank-style handle that opens the window outward. Because they pivot from the top, awning windows swing upwards when opened.

balloon framing – a form of construction platform consisting of wall studs that span from foundation to eave, with a floor platform suspended within the frame.

bent – a cross-sectional timber frame wall; bents run perpendicular to the ridge of the roof, while walls run parallel to the ridge.

bubble diagram – an architectural drawing that uses circles to represent spaces within a home; circles are used to represent functions of spaces as well and are drawn adjacent to each other to represent space-uses that require proximity, such as a pantry and a kitchen.

building code – regulations adopted by a state/province, city, town or municipality which address the minimum acceptable levels of safety for structures to ensure the health and welfare of the public.

building section – a construction document; a two-dimensional drawing or view that slices through the building vertically and shows how the walls, floors, ceilings and roof work together structurally.

building shell – the walls and roof including windows and doors; shell materials are assigned a thermal-resistance value, referred to as the R-value. See also: R-value.

building system – systems include electricity for lights, outlets and appliances; water for sinks and showers; and furnaces or boilers for supplying heat.

CAD – computer aided design. CAD software is used for generating design drawings for print or three-dimensional computer models.

casement window – a type of window that is operated with a crank-style handle that opens the window outward. Because they pivot from the left or right, casement windows swing to the side when opened.

collar tie frame – a style of frame that has no internal posts and is appropriate for narrower spaces. Collar ties are pieces of timber that span horizontally between rafter pairs.

common purlin frame – a general category of frame; includes any frame type that consists of a series of purlins bearing on principal rafters and often spanning from bent to bent at uniform intervals. See also: purlin, common rafter frame.

common rafter frame – a general category of frame; includes any frame type that consists of roof rafters that span from the eave walls to a ridge beam, principal purlin or opposing rafters. See also: purlin, common rafter frame.

cruck frame – a frame type that uses matched, naturally curved timbers that are used to form bents resulting in an "organic" look; one of the most ancient frame types known.

double-pane window – a more thermally efficient glass consisting of two panes, ½" (about 1 cm) apart and typically filled with an inert gas such as argon.

double-hung window – a window style consisting of two operable sashes of equal size, with one placed over the other. The top of the lower sash and the bottom of the upper sash line up vertically to form a weather-tight seal.

dovetail – a joint that relies on the mechanical properties of interlocking tapered wood pieces to hold the two ends together. See also: joinery/joints.

easement – a limited right for use of a private property by an outside party, such as the right of a municipality to run gas or electrical lines under or over a residential property; an easement does not give the holder a right of "possession" for the property, only a right of use.

eave – the lower, horizontal edge of a roof that overhangs the wall.

eave line – (see: eave)

eave plate – where the roof joins the walls.

eave posts – posts that are part of the eave walls.

eave wall – the wall beneath a roof eave upon which the roof rests.

elevations – the term for a type construction document that provides detailed design drawing showing the exterior views of the home.

energy recovery ventilator (ERV) – an air exchanger which transfers only the heat and humidity of outgoing exhaust air to the incoming fresh air.

expanded polystyrene (EPS) – a rigid foam insulation made of small beads of styrene foam that is formed into boards; used in the manufacture of a type of SIP.

extruded polystyrene (XPS) – a rigid foam insulation made of styrene foam forced through molds to create panels; used in the construction of a type of SIP.

floor plans – a type of construction drawing that shows a two-dimensional view of the home, as if it were sliced several feet above the floor height, looking down.

foundation plans – a type of construction document that represents a horizontal slice of the foundation system.

foyer – a room in a building which is used for entry from the outside; an entrance hall.

frame drawings – a construction document that provides details about frame dimensions as well as all required references needed for the erection of the frame; these refer to both timber framed portions and conventionally framed portions of the building.

frost wall – a foundation wall extending below the regional frost line but not deep enough to create a basement.

furring strip – a small dimension piece of lumber (1"x 3" or 2.5 cm x 7.6 cm for example) used to create space between a finishing material and the substrate material.

gable – the triangular shaped portion of a wall between the two sloping lines of a roof.

grade – a term used to define the slope of land.

gravity system – a septic system which relies primarily on gravity to move waste from the house fixtures to the septic field or city sewage connection.

great room – a living room or gathering place that has cathedral ceilings.

half-lap – a joint that relies upon removing equal and opposing amounts of wood from two opposing timbers; it can be pegged or rely solely on the mechanical properties of the form of the joint.

hammer beam truss frame – a common purlin frame style that exerts great pressure onto the eave posts with the inclusion of two hammer beams; similar to a queen post truss frame, with ties that break the post. See also: queen post truss frame, post.

heat pump – a system for cooling the home that works by either by removing the heat from inside and expelling it outside (air source), or by transferring the heat to and from the ground at a depth where the earth has a constant temperature (ground source).

heat recovery ventilator (HRV) – an air exchanger which transfers the heat of outgoing exhaust air to the incoming fresh air.

hydronic system – a heating system that can include baseboard radiators or under-floor radiant systems; requires a boiler to heat water that can move through the system and radiate heat to the house.

insulated concrete form (ICF) – an insulating method using expanded polystyrene that is formed into interlocking blocks and used to form both sides of a concrete foundation wall's formwork. Once the concrete is poured the forms remain in place providing superior insulation for below-grade applications.

joinery/joints – fasteners that hold timber pieces together, like nails and screws used in conventional framing, or wooden joinery, like pegs used in timber framing.

king post bent frame – a common purlin frame style using vertical timbers called king posts, which span from floor to ridge and are placed at regular intervals along the length of the ridge; diagonal braces from the king post to the rafters create the truss that helps support the roof load.

king post truss frame – a common purlin frame style similar to the king post bent truss, but with the addition of a tie that transfers the weight-bearing load to the eave posts, eliminating the king post from below the tie.

low-emissivity (low-e) – a coating applied to window glass that reduces heat loss or gain through the window; the lower the emissivity rating, the better.

massing – the process of defining physical volumes and their forms.

mortise and tenon – joints mechanically fastening two separate timbers; the basis of timber frame joinery. See also: joinery/joints.

mound system – a septic system that has a raised drain field because of inadequate or poor soil.

mudroom – an area between the formal house and the outdoors to shed, store and possibly wash outdoor garments.

nave – the central bay in an aisled frame. See also: aisled frame.

OSB – (see: oriented strand board)

oriented strand board (OSB) – a dense type of sheathing manufactured by pressing wood chips saturated with an adhesive into sheets similar to plywood in dimensions; used in the manufacture of some types of SIPs. See also: structural insulated panels.

panel drawing – a type of construction document that illustrates the layout of panels on the structure and correlates to a panel numbering system.

platform framing – a modern and conventional construction method that involves erecting wall studs that span from the first floor to the underside of the second floor, then building a platform at the top to facilitate the building of the next story (and successive stories).

porte-cochere – a covered entrance designed to shelter both people and vehicles.

post – a vertical load-bearing structural element in framing.

post and beam – framework based solely on upright posts and horizontal beams.

pre-design – the phase of planning that involves programming as well as the identification of applicable codes, regulations and zoning issues, and the consideration and preliminary selection of building systems (such as heating/cooling and electricity). Additionally, the scope of the project, consultant and contractor responsibilities, project phasing, project budget, budget contingencies and schedule should be established in this phase. See also: programming.

pressure distribution system – a type of septic system that uses a pump in lieu of gravity to remove waste.

principal purlins – longitudinal roof frame timbers that are used between the eave plates and the ridge, supported by internal posts. See also: eave plate, purlin, ridge.

principal rafters – the main rafter pairs of a bent, which typically support the common purlins.

programming – the process of creating a formal outline; includes architectural style preferences, desired site characteristics, required spatial relationships and important functional adjacencies, details about site and environmental conditions, room organization, spaces by story, and approximate square footage. See also: building systems, pre-design, regulations, zoning.

project schedule – typically performed by the contractor, who identifies the primary construction elements or work structures by trade or building system, and assigns start and completion dates.

purlin – a longitudinal roof timber used between principal rafters to carry the roof decking or SIPs. See also: structural insulated panel.

queen post bent frame – a type of common purlin frame that uses vertical "queen posts" that span from the floor to part-way along the length of the principal rafters, forming aisles along the eave walls. The queen posts are connected to one another with ties.

queen post truss frame – similar to the queen post bent frame, but with the addition of a tie that breaks the queen posts and transfers the loads to the eave posts, removing the internal posts that affect the floor plan in the queen post bent frame.

R-value – the value assigned to a material's ability to resist the flow of heat, with larger numbers indicating greater resistance.

radiant floor heating system – a heating system that delivers heat from the boiler through flexible piping either directly under the floor system or embedded in the floor system.

rafter – the sloped timbers that support the roof decking and/or SIPs, used to help carry the roof load. See also: structural insulated panel.

ridge – the upper edge of a roof formed by meeting rafter pairs, a ridge beam or a ridge purlin. The ridge is at the top of the triangle formed by trusses or where the rafter pairs meet. See also: purlin, rafter, ridge beam, ridge purlin, truss.

ridge beam – a horizontal timber that supports the rafter pairs and forms the ridge of the roof. See also: rafter, ridge.

ridge beam frame – a frame type that uses rafters that span from the eave plates to a central ridge beam or beams (depending on the length of the building). Ridge beam frames have posts located centrally to support the ridge beams and that must be considered in the floor plan layout.

ridge purlin – a horizontal timber that spans between the principal rafter pairs of a bent and forms the ridge of the roof. See also: bent, purlin, rafter, ridge, ridge beam.

roof eave – the lower, horizontal edges of the roof that typically overhang the walls.

sand filter – a special type of sand media used to augment or replace inadequate soil in the drain field of a septic system.

sash – the frame of the window that the glass is set into.

scarf joint – a form of joint resisting movement in more than one axial direction held fast with keys, wedges or pegs. See also: joinery/joints.

schedules – documents that provide detailed information about parts of construction documents and drawings; schedules are referenced in the construction documents and prevent drawings from being cluttered by extensive text and explanation.

schematics/schematic design – architectural drawings that are preliminary rough sketches intended to convey the breadth or scope of a portion of the design.

scissor truss frame – a type of common purlin frame that uses diagonal, crossing ties that span from eave posts to the opposite rafters; the use of truss struts supports loads efficiently and allows for a greater span than rafters alone.

sections and details – construction documents that are typically large-scale drawings in horizontal or vertical cross-section, and show specific instances of construction within the building.

septic system – an on-site waste removal system, usually buried in the ground, used if municipal waste water removal is not provided; a small-scale sewage treatment system.

setbacks – distances between the property line and where construction can occur on the property, imposed by zoning regulations; for example, the required minimum distances from the street front and neighboring properties.

sill – most commonly refers to the canted horizontal piece below a window or door opening, intended to shed water away from the opening.

SIP – (see: structural insulated panel)

sliding window - a window style consisting of two operable sashes of equal size, placed side by side with the left and opposing right side of the sashes forming a weather tight seal.

slope analysis – the surveying and plotting of the grade of a particular piece of land in relation to its elevation.

soffit – the underside of an eave overhang.

specialty regulations – restrictions or codes enforced by municipalities or counties that refer to specific local or regional issues and can vary from jurisdiction to jurisdiction.

spline joint – a long thin piece of wood that fits into corresponding mortises, used to fasten timbers. See also: joinery/joints.

structural insulated panel (SIP) – a manufactured panel used in a building shell that offers exceptional durability, superior insulation and strength. See also: expanded polystyrene; extruded polystyrene; oriented strand board; R-value.

sub-floor – flooring (often plywood or oriented strand board) that spans the floor joists providing rigidity as well as a surface to fasten a finish flooring material to. See also: oriented strand board.

tapered shoulder joints – weight-bearing joints used to connect horizontal timbers and vertical posts with mortise and tenon joints by providing a load-bearing surface that allows the horizontal timber to transfer its load to the post. See also: joinery/joints.

tenon – (see: mortise and tenon)

tie – a horizontal beam placed part of the way down the rafters, creating a smaller triangle. Ties are used to strengthen roofs and make the rafters more rigid.

timber frame – a traditional form of construction using sawn or hewn wood that is characteristically left exposed; joined by mortise and tenon joinery. See also: joinery/joints, mortise and tenon.

topography – the relief or terrain of a building site surface; the identification of specific landforms, vegetative and human-made features.

triple-pane window - a more thermally efficient glass consisting of three panes ½" (about 1 cm) apart; the space between is most often filled with an inert gas such as argon.

truss – a frame based on the strengths of joinery and the rigidity of the triangular form.

urethane insulation (in SIPs) – foam that is injected into a mold which then expands and adheres to the skins of SIPs. See also: structural insulated panel.

wattle and daub – a traditional infill between timbers, where thin planks were woven with twigs then plastered with a clay/straw mixture.

zoning – land-use regulations that prevent new developments from negatively affecting established land-uses and segregate incompatible uses, such as industrial and residential.

DESIGN WORKBOOK

by Jeremy Bonin

$14.95

INCLUDES:
Needs Analysis | Budget Plans | Design Ideas | Reference Material | Building Timelines | And More!

ORDER NOW!
1 800 636 0993   timber@davisframe.com
www.davisframe.com   www.classicbarnhomes.com

DAVIS FRAME CO.
AUTHENTIC CUSTOM TIMBER FRAME HOMES